Texts in Computational Science and Engineering

17

More information about this series at http://www.springer.com/series/5151

Bertil Gustafsson

Scientific Computing

A Historical Perspective

 Springer

Bertil Gustafsson
Department of Information Technology
Uppsala University
Uppsala, Sweden

ISSN 1611-0994 ISSN 2197-179X (electronic)
Texts in Computational Science and Engineering
ISBN 978-3-030-09915-2 ISBN 978-3-319-69847-2 (eBook)
https://doi.org/10.1007/978-3-319-69847-2

Mathematics Subject Classification (2010): 65-XX, 68-XX, 01-XX

Cover illustration: Figure 2.1 showing the Babylonian clay tablet. (The clay tablet YBC7289. From
the Yale Babylonian Collection, with the assistance and permission of William Hallo, Curator, and
Ulla Kasten, Associate Curator. Photo by Bill Casselman, http://www.math.ubc.ca/_cass/Euclid/ybc/ybc.
html.)

This Springer imprint is published by the registered company Springer Nature Switzerland AG
The registered company address is: Gewerbestrasse 11, 6330 Cham, Switzerland

Preface

Science and technology are traditionally based on theory and experiments, but nowadays scientific computing is an established third branch of strategic importance. It is based on mathematical models, some of which were developed many centuries ago. Most of these models are in the form of different kinds of equations that must be solved, and in the ideal case, one can find the solutions by analytic means in the form of an explicit mathematical form, as for example a function $f(x, y)$ that can be easily evaluated for any value of the independent variables x and y. However, in most realistic cases this is not possible, and numerical methods must be used to find accurate approximations of the solutions to the mathematical model. The concept of scientific computing usually means the whole procedure including analysis of the mathematical model, development and analysis of a numerical method, programming the resulting algorithm, and finally running it on a computer. Today, there are many systems that include the whole chain and require only specifying the physical parameters for the problem to be solved numerically. However, there are always new challenges with new types of mathematical models that require new numerical methods.

This book is about some of the most significant problem areas and the history of the process leading to efficient numerical methods. Mathematics in the old days was very closely connected to astronomy, which was always the major source for new mathematical problems up to the nineteenth century. According to Galileo Galilei (1564–1642), "the book of nature is written in the language of mathematics," and this was particularly true for astronomy. As we shall see, many of the most famous mathematicians in the past could as well have been called astronomers. For example, perhaps the most famous mathematician of all time, Carl Friedrich Gauss (1777–1855), was the director of the German astronomical observatory *Göttingen Observatory* for 18 years.

Today, there is a clear distinction between pure mathematics and applied mathematics with scientific computing quite far from pure mathematics. This was not the case in the old days. Mathematicians found new models for various physical problems, but they also carried out the necessary computations to find the unknown numbers they were looking for. A typical example was to predict the future location of a certain planet based on a few available observations.

The content of this book is divided into four parts. The first is about the very early mathematical/numerical achievements made by the Babylonians and the Greeks. After that not much happened until the seventeenth century when Newton and others developed new mathematics, and the second part is about the development during the centuries until the Second World War. Just at the end of the war, scientific computing took a giant step forward with the construction of electronic computers. Now numerical methods that earlier led to computations of impossible size could be implemented on these computers with results obtained after a few hours. In this new situation, the development of new methods became much more interesting, and the existence of electronic computers provided a strong boost for numerical analysis and scientific computing. The third part of the book is about this postwar period until the end of the 1950s. Around that time, scientific computing became a third scientific method in addition to the traditional branches in theory and experiments. The fourth part of the book covers the period until the present time.

The major numerical methods are traced back to their origin and to the people who invented them, as well as to the origin of important techniques for analysis of the methods. There is also short presentations of some of the mathematicians who played a key role in this process. There is certainly not a complete description of the whole story behind all methods, but rather an attempt to catch the key steps without going into a complete mathematical derivation. There is also a very brief presentation of the development of electronic computers, particularly the early ones.

Differential equations are the dominating type of mathematical models for almost every branch of science and engineering, and there are several different principles that are used when developing numerical methods for the solution of these problems. Therefore, differential equations are given some extra space in this book.

One difficulty when going back in time is that the original articles are not always easy to read and understand. During the active period from the sixteenth to the nineteenth century, Latin was a common language for mathematical texts. Another major difficulty was the way mathematics was described. Many symbols used today were not introduced at that time, and numerical methods were described in words in quite a lengthy and complicated way. In fact, in some cases there is still a certain uncertainty about the exact meaning of the description, and a consequence of this may be that the real inventor of a certain method is not known for sure.

We are aware of the fact that the whole area of scientific computing is not covered in this book. There are many techniques and methods that are left out or just briefly mentioned, and there are many mathematicians who could have been included but are not. The book is intended to give the big picture and the historical development behind the rise of scientific computing as a new scientific branch.

The content of this book is based on a large number of sources in addition to the original books and articles in the reference list. It is impossible to give all of these sources, in particular for some of the notes concerning important mathematicians, but as far as possible, correctness has been checked.

Hopefully, the book will be a valuable resource for all students and professionals interested in the history of numerical analysis and computing and for a broader readership alike.

Uppsala, Sweden   Bertil Gustafsson

Acknowledgments

This book was written after my retirement from the chair at the Division of Scientific Computing at Uppsala University. The Department of Information Technology has still provided full access to the necessary infrastructure, which is much appreciated.

Martin Peters and Ruth Allewelt at Springer have been helpful, in particular when it comes to copyright issues. Ann Kostant corrected language deficiencies and misprints, and I am impressed by her ability to find so many errors and language details that required correction.

Finally, I would like to thank my wife Margareta for accepting and supporting my full engagement in writing still another book.

Contents

Chapter 1
Scientific Computing: An Introduction

More than two and a half millenniums ago, the Greek mathematician Pythagoras came up with a mathematical model for the relation between the sides of triangles with a 90-degree corner.

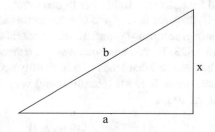

A straight angle triangle

The model is the famous Pythagorean theorem

$$x^2 + a^2 = b^2.$$

The lengths of sides a and b are known, and we want to know the length of x. The problem is that the model is an equation involving the unknown x, and it is not an explicit expression giving the value of x in terms of a and b. However, in this case the equation can be easily solved, and we have

$$x = \pm\sqrt{b^2 - a^2},$$

where the minus sign is of no interest here. From a mathematical point of view the problem can be considered as solved since we have an explicit expression for x. However, in order to get a number, the square root must be computed. If $a = 7$

and $b = 8$, then $x = \sqrt{15}$, but how is this number obtained? Not so long ago, one had to look it up in a table of square roots, but nowadays it is easily achieved by a few clicks on the simplest pocket computer or smart phone, and the number 3.8730 shows up. But this number is calculated by an iterative numerical algorithm that was already invented by the Greek mathematician Heron of Alexandria in the first century A.C. and later generalized to other problems by Newton as we shall see in Sect. 3.1.2. The point to be made here is that the simplest mathematical model like the Pythagorean theorem already requires a numerical algorithm in order to produce any useful results.

For the more general equation

$$x^2 + ax + b = 0$$

there is also a simple formula for the solutions:

$$x = -\frac{a}{2} \pm \sqrt{\frac{a^2}{4} - b},$$

which can be evaluated for a given a and b. For higher degree equations containing x^3- and x^4-terms as well, there are also explicit formulas for the solutions. However, if there are x^5 or higher degree terms present, no such general formulas exist, which was proved by Abel in 1823. In this case *numerical methods* must be used, i.e., some procedure must be invented that leads to a hopefully accurate approximation of x. Very effective such methods exist, actually found very early by Newton. So, for general polynomial equations

$$x^n + a_{n-1}x^{n-1} + \ldots + a_0 = 0,$$

the solution x can be found by analytical means if $n \leq 4$, but some numerical procedure must be used to evaluate the mathematical expressions involved. For $n \geq 5$ an approximate solution must be found by applying a numerical method directly to the equation.

With Newton and Leibniz came the introduction of derivatives when dealing with functions $u(t)$. If a body located at u_0 when $t = 0$ is moving with constant speed a, the mathematical model connecting the position $u(t)$ and speed is

$$\frac{du}{dt} = a,$$

$$u(0) = u_0.$$

This is an *initial value problem* involving a differential equation and an initial condition. Here we are looking for a solution in the form of a function $u(t)$, and it is easily found as

$$u(t) = u_0 + at, \tag{1.1}$$

which is an explicit expression that can evaluated for any value of t. As long as we are dealing with linear differential equations with constant coefficients

$$\frac{d^n u}{dt^n} + a_{n-1}\frac{d^{n-1}u}{dt^{n-1}} + \ldots + a_0 u = 0,$$

it is possible to write down the solution in an explicit form as a function of t. However, when the coefficients depend on t, we are in trouble, because there is in general no explicit form for the solution. It is even worse when the differential equation is nonlinear, for example in the common case where $a_k = a_k(t, u)$. Then it is not even clear if a solution exists, and if it does, there is no way to find it by analytical means. Even the sharpest mathematician, with all the mathematical theory available, does not stand a chance when it comes to solving most nonlinear differential equations. In fact, this should not be a surprise. We know already that a relatively simple problem like finding the roots of a fifth degree polynomial by analytical means is impossible, which has been proven theoretically. However, there are very effective numerical methods based on discretization, where the derivatives are substituted with finite difference formulas such that the solution can be computed at gridpoints $t_j = j\,\Delta t$ to very high accuracy.

Even more complicated problems arise when going to partial differential equations where the functions we are looking for depend on two or more variables. The great Swiss mathematician Leonhard Euler developed the mathematical model for fluid dynamics in the eighteenth century. This consists of a system of nonlinear partial differential equations connecting velocity, pressure and density. There is of course no explicit form of the solution, and in fact these equations still represent a great challenge to the sharpest minds who are trying to find efficient and robust numerical methods to find an approximate solution to them.

The development and analysis of numerical methods for finding solutions that satisfy mathematical models is called numerical analysis or computational mathematics, or scientific computing; the precise definition of these concepts varies. Applied mathematics is a wider concept than any of these three. It is usually interpreted as the general procedure of developing mathematical models and methods for solving problems arising from other scientific disciplines such as physics, biology, and engineering. Computational mathematics and scientific computing are concerned with *computing*, i.e., the procedure of actually producing numbers and graphs as solutions to problems defined by mathematical models.

Mathematics was used several thousand years ago, most notably by the Babylonians, the Greeks, and the Chinese. One of the main driving forces was the astronomical observations and the desire to understand the movements of the stars. The mathematics that was developed had a truly applied character, since it was designed to understand physical processes and to perform real computations. However, not much happened until the seventeenth century when Newton and others started a new era. The following centuries were strong periods for mathematics, but it was still driven mainly by physicists and astronomers. And then came a new start for computation.

There are few areas in science that got such a dramatic boost within a short period of time as did scientific computing. The introduction of electronic computers in the middle of the 1940s changed the situation completely. Suddenly, it was worthwhile constructing numerical methods even if the implementation required millions of basic arithmetic operations. Within a short time, scientific computing became a third scientific tool as a complement to the classical tools of theory and experiments.

We shall also see that the 1940s became an important decade from another point of view. The second world war required large computational efforts for weapon developments, and in particular for the Manhattan Project in the United States for developing the atomic bomb. Many of the sharpest minds became engaged in these efforts. It is also striking how Hitler's behavior, forcing a large number of Jewish mathematicians to flee Europe to the United States, had such a strong impact. Many of these made fundamental contributions to the war effort, and their work was the start of the very fast development of Scientific Computing.

In this book we give an overview of the development over more than two millenniums, with the primary emphasis on the development of numerical methods and the historical process behind them.

There are other books that contain a survey of numerical methods from a historical point of view. Two of them are:

Herman H. Goldstine: *A history of numerical analysis. From the sixteenth through the nineteenth century* [72].

Jean-Luc Chabert et al.: *A history of algorithms. From the Pebble to the Microchip* [20].

A comprehensive survey of the field is found in a series of articles collected in the book:

Björn Engquist (editor): *Encyclopedia of Applied and Computational Mathematics* [48].

Chapter 2
Computation Far Back in Time

Today we usually make a distinction between mathematics and applied mathematics, and the big challenges in mathematics have a truly abstract character. A typical example of such a challenge is the proof of the Riemann hypothesis which was formulated by Bernhard Riemann in 1859. It states that the nontrivial zeros of the Riemann function

$$\zeta(s) = \sum_{n=1}^{\infty} \frac{1}{n^s}$$

are located at the critical line $1/2 + ir$ in the complex plane, where r is real. Actually, it has been proved that the first 10 trillion (10^{13}) zeros are located there, but this is of course no proof of the general statement. The hypothesis is connected to the distribution of prime numbers, but its proof does not really contribute much practical value. Still, one can be sure that the one who finally proves it will be the most famous mathematician for a long time and awarded all sorts of prizes. Indeed, since the year 2000 there is a one-million dollar prize offered by the Clay Mathematics Institute to the person who proves it.

The zeros can of course be computed by numerical methods, and in this way there is a chance that a number s_0 could be found outside the critical line such that $\zeta(s_0) = 0$. However, the person who finds such a zero by using a computer does not receive the prize. In the old days, the situation was very different. One of the most famous mathematicians of all time, Archimedes, had a very broad range of interests and can be called mathematician, physicist, astronomer, engineer and inventor. At least until the eighteenth century most mathematicians were driven by various applications, with astronomy being the most important. It was the driving force for the development of many basic numerical methods that are still used. Another field was geometry, where many basic questions had to be answered. In this and the next chapter we shall describe some of the most significant developments

© Springer International Publishing AG, part of Springer Nature 2018
B. Gustafsson, *Scientific Computing*, Texts in Computational Science
and Engineering 17, https://doi.org/10.1007/978-3-319-69847-2_2

that took place from more than 2000 years ago until the middle of the twentieth century.

It is difficult to know what kind of mathematics and computations was done several thousand years ago. Clearly, geometrical problems fascinated many sharp minds. The Pythagorean theorem concerning right angle triangles is definitely one of the most well known results in geometry. It was explicitly presented and proved by the Greek mathematician and philosopher Pythagoras (580 BC–495 BC), but the result was already known by the Babylonians more than a millennium earlier. The problem, mentioned in the introduction, leads to the need for computing the square root of real numbers, which became a major challenge. Another known fact long ago was that the ratio between the circumference of a circle and its diameter is a constant, and the problem was to find this constant. Both of these problems lead to the concept of iteration, i.e., a sequence of increasingly accurate approximations are generated by some algorithm. In this chapter we shall describe early developments of such methods. Note that we limit ourselves to the part of mathematics which is related to numerical methods and computation.

2.1   The Babylonians

The earliest documented mathematics is from the Babylonian time covering almost one and a half millennium starting around 1800 BC. They used clay tablets to write their results, and many of these tables have been found. The basis for the notation of numbers was 60, and this is the reason for the still remaining partition of the hour into 60 min with 60 s to 1 min. They did know about the Pythagorean theorem, which can be seen on the famous clay tablet YBC 7289 from the Yale Babylonian Collection, see Fig. 2.1.

Fig. 2.1 The clay tablet YBC 7289 (From the Yale Babylonian Collection, with the assistance and permission of William Hallo, Curator, and Ulla Kasten, Associate Curator. Photo by Bill Casselman, http://www.math. ubc.ca/~cass/Euclid/ybc/ybc. html)

The use of 60 as the basis for the representation of numbers requires 60 different symbols when writing a number. The symbols on the tablet have been translated into ordinary decimal numbers in the picture, and the number on the horizontal diagonal is

$$x = 1 + 24 \cdot 60^{-1} + 51 \cdot 60^{-2} + 10 \cdot 60^{-3},$$

which to 7 decimal digits is $x = 1.4142130$. This is an amazingly accurate value of $\sqrt{2} = 1.4142136$. The tablet also shows that with the side 30, the diagonal is $42 + 25/60 + 35/3600$, which is $30 \cdot 1.4142130 \approx 30\sqrt{2}$. The question now is: how was the approximate value of $\sqrt{2}$ computed?

Actually, nobody knows. Perhaps the most likely methods were based on extensive tables of squares that certainly were available. One way would then have been to take an average of an initial lower bound a and an upper bound b. If the square of $(a + b)/2$ were larger than 2, then replace b by this better upper bound; if the square were less than 2, then replace a by this better lower bound.

In those days there was not much mathematical notation available, and algorithms were described largely in words. This continued to be the case until the eighteenth century AD. However, in modern notation the method above for square roots can be expressed as follows.

Assume that there are two approximations such that $a_0^2 < 2$ and $b_0^2 > 2$. Then the algorithm can be formulated as

$$if \; \left(\frac{a_n + b_n}{2}\right)^2 > 2 \quad then \quad a_{n+1} = a_n, \; b_{n+1} = \frac{a_n + b_n}{2},$$

$$if \; \left(\frac{a_n + b_n}{2}\right)^2 < 2 \quad then \quad b_{n+1} = b_n, \; a_{n+1} = \frac{a_n + b_n}{2},$$

$$n = 0, 1, \ldots.$$

This may be one of the first iterative methods that was ever used for computation. Convergence may be slow, but it works.

Algorithms based on iteration are very common in computational mathematics, and a theoretical analysis of their convergence properties is a central part of numerical analysis today. In a way this may be surprising, since modern computers are so fast 4000 years later. What took mathematicians weeks in Mesopotamia to compute takes a very small fraction of a second on a computer today. However, mankind seems to be pushing for larger and larger problems with billions of unknowns coupled to one another so that the convergence rate becomes important.

Some authors have suggested that the Babylonians used another more sophisticated method to compute square roots. Let us generalize a little, and discuss the general problem of finding $x = \sqrt{a}$, where a is a real positive number. Let x_0 be an initial guess with an error ε such that

$$a = (x_0 + \varepsilon)^2 = x_0^2 + 2\varepsilon x_0 + \varepsilon^2. \tag{2.1}$$

We would like to solve for ε, but of course this leads nowhere, since it requires solving a quadratic equation, again leading to computing a square root. However, if x is not very small, and x_0 is not too far from the true solution, the last term ε^2 is small compared to the others. By neglecting it, we get

$$\varepsilon \approx \frac{a - x_0^2}{2x_0},$$

and it makes sense to define a new approximation:

$$x_1 = x_0 + \frac{a - x_0^2}{2x_0} = \frac{x_0}{2} + \frac{a}{2x_0}.$$

With this new and better value for $\sqrt{a}$, one can of course repeat the same procedure, and we get again an iterative method:

$$x_{n+1} = \frac{x_n}{2} + \frac{a}{2x_n}, \qquad n = 0, 1, \ldots \quad . \tag{2.2}$$

The method goes sometimes under the name of the *Babylonian method*, even if it is questionable whether the Babylonians used it. The first documented explicit presentation of the method was given by Heron of Alexandria (10–70), and it is therefore usually called *Heron's method*.

The method converges very quickly. Actually, it is quite easy to show that when we are in the neighborhood of the true value $\sqrt{a}$, then the number of correct decimals are doubled with each iteration. Heron's method can therefore be used to calculate by hand, which explains why it was used in former times.

We can look upon Heron's method from another point of view. Equation (2.1) is the true equation for the unknown ε, but the solution requires again a square root. In order to get an equation that we can solve, we make it linear, in this case by simply disregarding the quadratic term. This procedure is called *linearization*. The principle is simple: the equation for the correction is approximated by a linear equation. Under certain conditions this equation is close to the original as long as we are in the neighbourhood of the true solution. This principle of linearization is central and very important as a tool in computing, and we shall come back to it several times later in this book.

Some effort is still made to interpret what the Babylonians did in the past, and in particular what kind of mathematics they developed. In January 2016, Mathieu Ossendrijver at the Humbold University in Berlin published new results from his research in the history of science. The results were based on the content on 5 clay tablets kept at the British Museum in London regarding the movement of the planet Jupiter, where the fifth tablet was found in 2014.

The capital Babylon had its own god Marduk and Jupiter was his planet. Astrology was the basis for much thinking and planning in life, and the movement of Jupiter was therefore important to understand. Astronomers were able to measure

Fig. 2.2 Babylonian
computation of Jupiter's
position

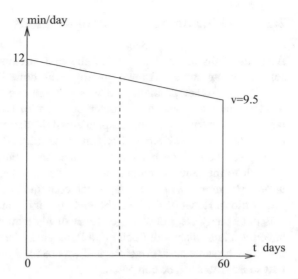

its velocity $v(t)$ across the sky at certain points in time, and the unit was in degrees
per day.

They didn't have access to any differential and integral calculus, so the formula

$$d = \int_{t_0}^{t_1} v(t)\, dt \qquad (2.3)$$

for the distance d covered during the time interval $[t_0,\ t_1]$ was not known. However,
if $v(t)$ is plotted in a graph as a function of t, the distance d is the area of the domain
under the curve, and this seems to have been known. The Babylonians used 60 as the
basis for representing numbers, and angles were expressed in degrees and minutes
just as we do today. The tablet can be interpreted as in Fig. 2.2.

With the velocity expressed in minutes at $t_0 = 0$ and $t_1 = 60$ days, they
approximated the velocity by an interpolating straight line and computed the area
under it as 645 min or equivalently, 10° and 45 min, which is the total movement of
the planet during this time interval. It seems that they were interested in computing
the point in time when it had moved halfway, i.e., finding the vertical line separating
the two domains with equal area. This is marked with a dashed line in the figure,
and it is located at $t = 28.3$ days which possibly was interpreted as the point in time
that the god Marduk had in mind.

From a computational point of view, the technique they used is quite interesting.
The integral is actually computed by using the trapezoidal rule that will be discussed
in Sect. 3.3. They didn't see it that way of course, but the basis for it is the
approximation of a function by a straight line, which they obviously did.

2.2   Archimedes and Iterative Methods

After the Babylonian time, one of the first attempts to use numerical methods
for computing some unknown quantity was done by Archimedes (287 BC–212
BC). Among other problems, he was interested in finding the relation between
the circumference and the diameter of a circle. It was known that the quotient
between the two is a constant independent of the diameter. In other words, he wanted
to find the numerical value of π (this notation had not been introduced yet). In
principle, it could have been found by measurements, but it would have been very
difficult to measure the circumference with any degree of accuracy. Archimedes
wrote *Measurements of a circle*, which contained an estimate of π as one of three
propositions. He worked with right angle triangles and the length of their sides, but
it is not exactly clear how he got some of his numerical results even if he would
have lived later than both Euclid and Pythagoras. Here we choose a version that is
based on the same basic principle, and we use the Pythagorean theorem and assume
that square roots can be computed.

We begin by drawing a circle and then an inscribed six-sided polygon with equal
sides as in Fig. 2.3.

Since the angle between the solid line legs of the triangle is 60°, all three sides
are equal with length r. Consequently, the total length of the inscribed polygon is $6r$,
and it is obviously shorter than the circumference of the circle. Archimedes could
conclude that π must be larger than 3.

To obtain better accuracy we double the number of sides in the inscribed polygon.
In order to compute the length of the new and shorter sides, triangles connected to
the new polygon are introduced as shown in Fig. 2.4. At this time the trigonometric
functions had not been invented, but the Pythagorean theorem can be applied to any
one of the triangles. Since π is independent of the radius, we assume $r = 1$ and

Fig. 2.3 Circle with
inscribed 6-sided polygon

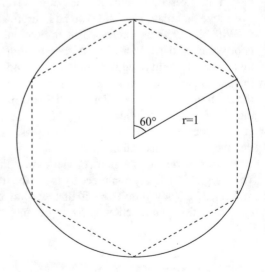

Fig. 2.4 Triangles associated
with the polygons

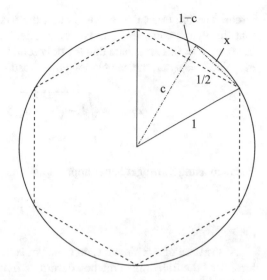

compute the new chord x. We know that the old chord is one, but for future use we
use the notation a.

The length c is found by

$$c^2 + (\frac{a}{2})^2 = 1,$$

i.e.,

$$c = \sqrt{1 - (a/2)^2}.$$

(The square root concept was not defined at this time, but we use it here for
convenience.) We next consider the smaller triangle, and compute the longest side
x. Again, by the Pythagorean theorem,

$$(a/2)^2 + (1 - c)^2 = x^2,$$

which has the solution

$$x = \sqrt{2 - \sqrt{4 - a^2}}.$$

For $a = 1$ corresponding to the figure, we get $x = 0.52$. Accordingly, the new and
better value of π is $6 \cdot 0.52 = 3.12$.

The procedure can now be continued by doubling the number of sides in the
polygon once more, and then applying the same type of calculation on the new
triangle. At this point we can make an important observation. The computation
of x above is independent of the angle at the triangle corner at the center of the

circle. The only fact that counts is that the triangles have one straight angle such that the Pythagorean theorem can be applied. Therefore we introduce an iteration index n, $n = 0, 1, \ldots$, such that the polygon P_n has $m_n = 3 \cdot 2^{n+1}$ sides with length a_n. With $x = a_{n+1}$, the calculation above results in the formula

$$a_{n+1} = \sqrt{2 - \sqrt{4 - a_n^2}}, \quad n = 0, 1, \ldots,$$

$$a_0 = 1.$$

(2.4)

For increasing n, we get better approximations:

$$\pi_n = 3 \cdot 2^n a_n$$

(2.5)

of π.

At the time of Archimedes, no tables of square roots were available, but he may have used the Babylonian method described in the previous section. No matter what method he used, the computational work must have been very cumbersome, and he stopped at $n = 4$ corresponding to a 96-side polygon. This gives the value 3.1408, which is $3 + 10/71$ obtained by Archimedes.

The algorithm (2.4) is a nice example of an *iterative method* that is so typical for numerical methods of today. Starting out with an initial guess, one finds a formula that computes a new and better approximate value from the previous one. It is important to make sure that the sequence of numbers *converges* to a finite limit, which is the correct solution. In our case we know that the circumference of the circle is finite, which is then an upper bound for the sequence. However, one should be aware that this conclusion holds only if the calculation is made exactly. Our example is bound to give errors in the calculation, since we are dealing with square roots, and we cannot allow these errors to grow without bound when the iteration continues. The iteration must be *stable*. We shall come back later to this very important issue.

Even without rounding errors, there is always the uncertainty about the accuracy of the answer. Can we give an upper bound on the error that is not too pessimistic? In more complicated problems one cannot. However, in computing π it is indeed possible, and we shall show how it can be obtained.

Besides the inscribed polygon, we can circumscribe a polygon as shown in Fig. 2.5. The length of the circumscribed polygon is obviously larger than the circumference of the circle, and it can be computed in much the same way as for the inscribed polygon. However, by simultaneously involving both the inscribed and circumscribed polygons, and using relations between conforming triangles, one can derive simple formulas that describe one step of the iterative procedure in terms of the side-length a_n of the inscribed polygon and the side-length b_n of

Fig. 2.5 Inscribed and
circumscribed 6-sided
polygons

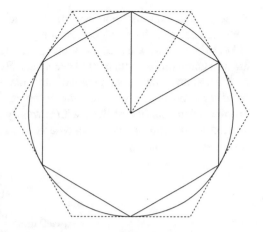

the circumscribed polygon. We omit the details here, and state the end result as
Archimedes found it:

$$b_{n+1} = \frac{a_n b_n}{a_n + b_n},$$
$$a_{n+1} = \sqrt{a_n b_{n+1}/2}.$$

(2.6)

For $n = 0$ we already know $a_0 = 1$, while b_0 is found as $b_0 = 2/\sqrt{3}$. With these
initial values, the iterative formula (2.6) can be used to find new approximations
to any degree of accuracy. Besides the lower bounds π_n already defined in the first
algorithm, we now have upper bounds

$$\hat{\pi}_n = 3 \cdot 2^n b_n,$$

(2.7)

such that

$$\pi_n < \pi < \hat{\pi}_n.$$

The iteration formula (2.6) is simpler compared to (2.4) in the sense that there is
only one square root computation. However, they both lead to the limit solution π.
Furthermore, in real life we must of course break the computation at some finite n,
and there will always be an error no matter how many iterative steps we can carry
out. Archimedes 96-sided polygons give the limits

$$3.1408 < \pi < 3.1429,$$

or, as Archimedes gave them

$$3 + 10/71 < \pi < 3 + 1/7.$$

Since π is an irrational number, the sequence of decimal digits never ends. Many new methods to achieve higher accuracy have been invented over the years. In the effort to break new world records, computations of even more correct decimal digits are still going on more than 2000 years later. Even if it is of little practical value, it is an illustration of the development of new iterative methods with faster convergence. In general, no matter how fast the computers become, there will always be new problems that cannot be solved with currently available methods.

Archimedes didn't have trigonometry at his disposal. With this tool available, the formulas for a_n and b_n are

$$a_n = 2\sin(\frac{180°}{m_n}),$$

$$b_n = 2\tan(\frac{180°}{m_n}),$$ (2.8)

where m_n is defined above.

We have emphasized the importance of the iteration principle that was used by Archimedes and illustrated here by the algebraic formula (2.6). However, there is another important concept that is illustrated as well by this example, and that is the principle of *discretization*. The continuous perimeter of the circle is substituted by a number of discrete points which are connected by straight lines. If the angle θ of the triangle corner at the center of the circle is taken as the independent variable and measured in degrees, we can think of the discretization having a stepsize

$$\Delta\theta = \frac{360}{6 \cdot 2^n}, \qquad n = 0, 1, \ldots$$

that tends to zero with increasing n. More than two millenniums after Archimedes, this principle is used again and again by modern computer algorithms.

Not much is known about Archimedes' life. Sicily was a colony of Greece at this time, and Archimedes lived in Syracuse, which was a very important city in the Mediterranean world. He was certainly one of the greatest mathematicians of all time, but he was also very much interested in astronomy, physics and engineering. On the mathematical side he introduced a new way of expressing numbers as exponentials, and proved the fundamental law $10^m 10^n = 10^{m+n}$.

There are many stories about him, some of them questionable. But almost certainly he was killed when the Romans invaded Syracuse. One story tells that a Roman soldier ordered Archimedes to follow him, but Archimedes was busy with some geometric problem and said "Do not disturb my circles". This made the soldier mad and he killed him with his sword. It is not clear if this is true, but whatever the reason for killing him, the famous general Marcellus didn't like this at all, since he knew about Archimedes' skills, and counted on the benefit of having him working for the Roman empire. More than 2000 years later, there was another person who didn't understand this wise thinking. Hitler's antisemitic actions forced

Archimedes (Erica Guilane
-Nachez/stock.adobe.com)

many sharp Jewish scientists to flee Germany and the surrounding countries. This certainly changed the scientific world map.

A large number of portraits from different sources has surfaced, and several sculptures have been created, many of them during the last centuries. They are all different, and there is little likelihood that any one resembles the true appearance. The portrait shown above is one of them.

2.3 Chinese Mathematics

In China mathematics was certainly a topic of interest, and several efforts to solve certain problems were made quite early. The problem is that documentation is hard to find. However, it seems clear that the first efforts were made about one millennium BC, also along with astronomy as one of the driving forces. One of the earliest complete documents is the *The Nine Chapters on the Mathematical Art* with unknown authors. It is a collection of problems and solutions representing a wide variety of applications over more than a millennium. Liu Hui came out with a commentary about it in 263, where large parts of it were interpreted and explained in some detail.

From a computational point of view, it contained the few problems that were of prime interest for many of the early mathematicians over several centuries, namely the computation of square roots and of π. The first one arose from the problem of finding the sides of a square with the area given, but also the third root became a problem when finding the sides of a cube with the volume given. The π-approximation 3.141014 was obtained by Liu Hui himself by using a 192-side polygon approximating the circle.

The Nine Chapters book also contained the quite advanced problem of finding the solution of a linear system of equations. By using a method essentially equivalent to Gaussian elimination, such a system was solved up to four unknowns.

Mathematics had a steady development during nearly a millennium after this book. From a numerical point of view, one can note the refined computation of π carried out by Zu Chongzhi (429–500) who obtained the quite accurate value 3.1415926. It is remarkable that this value of π was the most accurate one for almost a millennium forward in time.

A much later development was made by Li Yeh (1192–1279) who solved polynomial equations up to the 6th degree. It is not quite clear how he managed to do it, but it is assumed that the method is strongly related to Horner's scheme which is based on factoring the polynomial and finding the roots one at a time. Another type of problem was considered by Guo Shoujing (1231–1316) who introduced high-order differences connected to high-order interpolation.

Chinese mathematics had a quite strong period for more than a millennium AD, at least relative to the rest of the world. However, because of the lack of communications between the East and the West, the Chinese results were quite unknown in the rest of the world. During the first centuries AD, the communications improved resulting in a stronger western influence in Chinese mathematics. The Europeans took over as the prime inventors of newer developments in mathematics, not the least in the numerical parts of it. But when the Europeans started the development of mathematics on a much larger scale around the fifteenth century, it seems that they were not really aware of earlier Chinese achievements.

Chapter 3
The Centuries Before Computers

For a long time after the Greek era not much happened when it comes to computation. Certainly Chinese mathematicians were active. Some of them continued to compute π to higher accuracy, and in Sect. 3.5 we shall see that they were early developers of direct methods for linear systems of equations. However, their impact on the development of numerical methods for more general problems was not significant.

We have shown that methods for computing square roots existed, and it helped when solving many geometrical problems. Furthermore, if square roots can be computed, one can solve a full second-degree equation

$$ax^2 + bx + c = 0,$$

once the formula for the two roots are known. However, for a very long time no serious attempts were made to generalize Heron's square root algorithm to more general equations. But with the astronomical discoveries beginning in the sixteenth century with Kopernikus (1473–1543), Kepler (1571–1630) and others, a new computational momentum occurred.

However, these computations were limited to specific problems. No systematic efforts were made to develop general numerical methods until some of the greatest mathematicians took on this challenge beginning in the seventeenth century. In this chapter we will describe what happened from then on until the second world war.

There continues to be a discussion about the ranking of the greatest mathematicians of all time. Many ranking lists exist, and most likely, future lists will never converge to a unique one, not even for the top position. However, there are three

This chapter contains references to Wikimedia. Wikimedia Commons is a media file repository making available public domain and freely-licensed educational media content (images, sound and video clips) to all.

© Springer International Publishing AG, part of Springer Nature 2018
B. Gustafsson, *Scientific Computing*, Texts in Computational Science and Engineering 17, https://doi.org/10.1007/978-3-319-69847-2_3

Nikolaus Kopernikus
(Grafissimo/Getty
Images/iStock; 461241753)

names that occur on almost every list, and they are Newton, Euler, and Gauss
representing the seventeenth, eighteenth and nineteenth centuries respectively.
(More information about them will be given later in this book.) In contrast to
present day mathematicians, they all worked on a very broad range of problems,
also including the construction of mathematical models for problems in physics.
But there is also another observation to be made. They were not content with just
having a valid mathematical model for a certain physical process; they were also
interested in getting numerical results. This required developing numerical methods,
and it is striking how often these three names occur in the numerical literature. We
have the Newton method for solving nonlinear algebraic equations, Euler's method
for solving differential equations, and Gauss's method for solving large systems of
linear equations.

Until the beginning of the twentieth century, the development of numerical
methods was driven by the need for solving problems in various applied areas. At
first it was astronomy and various geodetic problems, then came problems in physics
and engineering. Numerical analysis did not exist as a topic by itself, and there were
hardly any courses at the universities. Great Britain was early with starting courses
given at the Computation Laboratory at the University of Edinburgh created at the
beginning of the century by Edmund Taylor Whittaker (1873–1956). He was also
the coauthor of what is probably the first book with *Numerical Mathematics* in its
title, [162] published in 1924. The book [148] by J.B. Scarborough published in the
USA in 1930 was called *Numerical Mathematical Analysis*. In a review 1932 by
W.R. Longley in the *Bulletin of the American Mathematical Society* it is said "... it
seems likely that the book will be most widely used for individual study without the
guidance of a teacher ... ", indicating the lack of university courses on the topic.

The real boost for numerical mathematics came in the middle of the 1940s as
a consequence of weapon development during World War II and the introduction
of electronic computers which happened almost simultaneously. In this chapter we
shall discuss the most significant numerical methods that were developed during
the centuries from Newton to that time. We shall see that in addition to the three
mathematicians mentioned above, many other famous names contributed to the

Johannes Kepler
(GeorgiosArt/Getty
Images/iStock; 177408395)

development of basic numerical methods that are still the basis for up-to-date algorithms.

3.1 Nonlinear Algebraic Equations

3.1.1 The Fixed Point Method

Kepler was an astronomer with an immense impact on future research in astronomy. But he also included numerical methods in his gigantic work *Epitome Astronomiae Copernicanae* published in the period 1617–1621. A central problem at the time was to determine the position x of a body moving around another body along an ellipse with eccentricity e at a given time t (x is actually expressed as an angle, and t is real time multiplied by a constant). This leads to *Kepler's equation*

$$x = t - e \sin x,$$

where t and e are given. This equation cannot be solved by analytical means, so Kepler used an iterative procedure that he described in words. Notation for a sequence of numbers $\{x_n\}$ was not used; instead at this time one typically used the notation $a, b, c, d, \ldots$ However, Kepler's method can be written in the form

$$x_{n+1} = t - e \sin x_n, \qquad n = 0, 1, \ldots.$$

In this way he could compute the position with quite good accuracy for a number of t-values. Actually, Kepler stopped the iteration after two steps, but the principle of iteration was introduced for the purpose of solving celestial problems described by a nonlinear algebraic equation.

If there is a value ξ such that a certain function $f(x)$ attains the same value, i.e., $\xi = f(\xi)$, then ξ is called a fixed point of f. Therefore Kepler's method applied to a general equation $x = f(x)$ was later given the name *fixed point method* or *fixed point iteration* with the general form

$$x_{n+1} = f(x_n), \qquad n = 0, 1, \dots .$$

If it converges as $n \to \infty$, the limit x_∞ clearly is a solution to the original equation.

When a new technique was invented in former times for the purpose of solving a specific problem, there was not much analysis, the technique was tried out immediately, and after heavy-handed calculations, one was satisfied if the result was reasonable. One reason for the lack of analysis was of course that there were not many tools available, even if differential and integral calculus became available after Newton. For iteration methods the basic question was whether or not the method converged as the number of iterations tends to infinity. However, this was a matter of theoretical interest only, since in practice very few iterations could be carried out. But it was of course important that the result after an extra iteration would be more accurate compared to the previous one.

For the fixed point iteration, a sufficient condition not only for the existence of a solution but also for convergence is the following. Assume that $f(x)$ is differentiable in an interval $[a, b]$ and such that the function values satisfy $a \leq f(x) \leq b$; then there exist exactly one fixed point ξ, and the iteration converges to ξ if

$$|\frac{df}{dx}(x)| \leq \alpha < 1 \text{ for all } x \text{ with } a \leq x \leq b.$$

It follows that the interval can be reduced such that it holds in a small neighborhood of $x = \xi$ if the iteration starts with an x_0 that is sufficiently close to ξ. Furthermore, the iteration certainly does not converge if $|df/dx| > 1$ at the point $x = \xi$.

Clearly a certain given equation $g(x) = 0$ can be rewritten in the form $x = f(x)$ in different ways. First, it may happen that the derivative condition is not satisfied at $x = \xi$, and even if it is, the convergence may be slow. The optimal convergence rate is obtained if $|df/dx|$ is as small as possible.

3.1.2   Newton Methods

It is no surprise that the mathematical giant Isaac Newton (1642–1727) really started the new era of computation. He had an usually unlucky start of his life, since he was born after the death of his father. When he was three, his mother remarried, and Isaac was left in the care of his grandmother. He thought that his mother did wrong when remarrying, and didn't like either his stepfather or his mother. He made a list of his sins up to the age of nineteen, and formulated one of them as "Threatening my father and mother Smith to burn them and the house over them"

(Newton biography-University of St Andrews), http://www-groups.dcs.st-and.ac.
uk/history/Biographies/Newton.html). When he was a teenager his mother, who by
then had lost also her second husband, tried to get her son to quit school and become
a farmer.

Isaac Newton (Portrait by
Godfrey Keller. 1689IMS
Bulletin. Institute of
Mathematical Statistics, IMS;
https://owpdb.mfo.de.
Oberwolfach Photo
Collection)

However, the master of the King's School at Grantham managed to get him back,
and his extraordinary intellectual skill became clear. At nineteen he was admitted
to Trinity College, Cambridge, and later he was elected Fellow of Trinity College.
That required him to be ordained as a priest. His scientific accomplishments became
a problem since it had to do with the physical laws of the universe, but after some
time he managed to escape this condition by special permission from Charles II. The
year was 1669, when Newton was appointed to the prestigious position as Lucasian
Professor of Mathematics at the University of Cambridge.

Newton was not a politician but he was elected as a member of the Parliament
of England representing Cambridge University. In 1703 he became President of the
Royal Society, and 2 years later he was knighted by Queen Anne.

During the last part of his life he became rather eccentric, and according to some
people, this was due to mercury poisoning as a result of his alchemical activities.
He died as a bachelor in 1727.

Newton was without doubt one of the greatest scientists of all time. The poet
Alexander Pope wrote in his epitaph:

Nature and nature's laws lay hid in night
God said "Let Newton be" and all was light.
(http://www.goodreads.com/quotes/168406-nature-and-nature-s-laws-lay-hid-in-night-
god-said.)

Newton's most well-known contribution to mathematics is in differential and integral calculus, with its first embryo included in *Philosophiae Naturalis Principia Mathematica* published in 1687. Gottfried Leibniz (1646–1716) developed the same type of calculus independently, and there was a bitter controversy about who was the first, which continued until Newton's death. However, as for numerical methods, the score clearly stands in Newton's favor.

As we have noted before, the mathematical problems at this time almost exclusively arose from problems in astronomy. For example, if a few observations of the position of a celestial body were known at certain points in time, how could the position be found for other points in time in which no observations had been made? This gave rise to the interpolation problem. Polynomials were a convenient choice for interpolating functions. The reason is that polynomials are easy to evaluate at any point by a sequence of simple additions and multiplications. We shall come back to the interpolation problem, but here we conclude that polynomials were a dominant class of functions during Newton's time.

One central problem was to find the roots of polynomials, in particular the real roots of polynomials with real coefficients. Even if explicit formulas for the roots exist for polynomials of degree up to four, it may be easier to find accurate approximations by some iterative method. Indeed, Newton invented a method by constructing a sequence of polynomials; and by neglecting higher-order terms, he was able to compute new and better approximations at each step. His results are included in the book *Method of Fluxions* which was not published until 1736. However, already in 1690 Joseph Raphson (1648–1715) had published his more general version of the Newton method. His method was based on the same ideas, but could be applied to the general equation $f(x) = 0$, where $f(x)$ is a real-valued function. Furthermore, it was presented and formulated differently. His formulation goes under the name Newton–Raphson method. This is still the most well known and most frequently used method for finding the zeros of a function. It may not always work in its original formulation, and several modified versions have been developed. However, this whole class of methods available today are usually referred to as *Newton type methods*.

We want to find a solution to the equation $f(x) = 0$, where $f(x)$ is a smooth function. Let x_n be an approximation of the true solution x^* such that $x^* = x_n + \varepsilon$. Then

$$0 = f(x_n + \varepsilon) = f(x_n) + \varepsilon f'(x_n) + \mathcal{O}(\varepsilon^2).$$

As above for Heron's method, we drop the ε^2-term, and obtain

$$\varepsilon \approx -\frac{f(x_n)}{f'(x_n)},$$

giving the original *Newton–Raphson method*

$$x_{n+1} = x_n - \frac{f(x_n)}{f'(x_n)}, \qquad n = 0, 1, \ldots. \tag{3.1}$$

(We note that this is exactly Heron's method for the equation $x^2 - a = 0$.) The method converges very fast as long as the derivative $f'(x_n)$ is well separated from zero. In fact, the convergence rate is easily derived in the following way. With $\varepsilon_n = x^* - x_n$ we have

$$0 = f(x^*) = f(x_n) + \varepsilon_n f'(x_n) + \frac{\varepsilon_n^2}{2} f''(\xi)$$

for some ξ near x^*, and after division by $f'(x_n)$

$$\varepsilon_n = -\frac{f(x_n)}{f'(x_n)} - \varepsilon_n^2 \frac{f''(\xi)}{2f'(x_n)} = x_{n+1} - x_n - \varepsilon_n^2 \frac{f''(\xi)}{2f'(x_n)} = -\varepsilon_{n+1} + \varepsilon_n - \varepsilon_n^2 \frac{f''(\xi)}{2f'(x_n)}.$$

If $f'(x_n)$ is bounded from below, this gives

$$|\varepsilon_{n+1}| \le \text{const} \cdot \varepsilon_n^2.$$

This is called *quadratic convergence*, which means that the number of correct decimal digits doubles with each iteration provided we are sufficiently near the solution.

The Newton–Raphson method can be generalized to systems of nonlinear equations. For compact notation we introduce the vectors

$$\mathbf{x} = \begin{bmatrix} x_1 \\ x_2 \\ \vdots \\ x_m \end{bmatrix}, \qquad \mathbf{f} = \begin{bmatrix} f_1 \\ f_2 \\ \vdots \\ f_m \end{bmatrix}.$$

Each element f_j is a function of all the variables x_j, $j = 1, 2, \ldots, m$, i.e., $\mathbf{f} = \mathbf{f}(\mathbf{x})$. We want to find the solution $\mathbf{x} = \mathbf{x}^*$ to

$$\mathbf{f}(\mathbf{x}) = \mathbf{0},$$

which is a system of m scalar equations for the m unknowns x_j, $j = 1, 2, \ldots, m$. Let $\mathbf{x}$ be an approximate solution, such that $\mathbf{x}^* = \mathbf{x} + \varepsilon$, where the vector norm $\|\varepsilon\|$ is small. For each element f_j there is a truncated expansion

$$f_j(\mathbf{x}^*) = f_j(\mathbf{x}) + \sum_{k=1}^{m} \frac{\partial f_j}{\partial x_k}(\mathbf{x})(\varepsilon) + \mathcal{O}(\|\varepsilon\|^2), \quad j = 1, 2, \ldots, m.$$

This expansion written in vector form is

$$0 = \mathbf{f}(\mathbf{x}^*) = \mathbf{f}(\mathbf{x}) + \mathbf{f}'(\mathbf{x})(\mathbf{x}^* - \mathbf{x}) + \mathcal{O}(\|\mathbf{x}^* - \mathbf{x}\|^2),$$

where $\mathbf{f}'$ is an $m \times m$ matrix called the *Jacobian* of the vector $\mathbf{f}$:

$$\mathbf{f}'(\mathbf{x}) = \begin{bmatrix} \frac{\partial f_1}{\partial x_1} & \frac{\partial f_1}{\partial x_2} & \cdots & \frac{\partial f_1}{\partial x_m} \\ \frac{\partial f_2}{\partial x_1} & \frac{\partial f_2}{\partial x_2} & \cdots & \frac{\partial f_2}{\partial x_m} \\ \vdots & & \ddots & \vdots \\ \frac{\partial f_m}{\partial x_1} & \frac{\partial f_m}{\partial x_2} & \cdots & \frac{\partial f_m}{\partial x_m} \end{bmatrix}.$$

The name of the matrix refers to the German mathematician Carl Jacobi (1804–1851) who made fundamental contributions to matrix theory.

When disregarding the squared terms, a new approximation $\tilde{\mathbf{x}}$ of $\mathbf{x}^*$ is defined by the linear system of equations

$$\mathbf{f}(\mathbf{x}) + \mathbf{f}'(\mathbf{x})(\tilde{\mathbf{x}} - \mathbf{x}) = \mathbf{0},$$

and the iteration formula becomes

$$\mathbf{x}^{(n+1)} = \mathbf{x}^{(n)} - \left(\mathbf{f}'(\mathbf{x}^{(n)})\right)^{-1}\mathbf{f}(\mathbf{x}^{(n)}), \quad n = 0, 1, \ldots.$$

This is the Newton–Raphson formula for nonlinear vector equations. The inverse $(\mathbf{f}')^{-1}$ is seldom computed explicitly when implementing the method. Instead one solves the system of equations

$$\mathbf{f}'(\mathbf{x}^{(n)})\varepsilon = -\mathbf{f}(\mathbf{x}^{(n)}) \tag{3.2}$$

and then computes $\mathbf{x}^{(n+1)} = \mathbf{x}^{(n)} + \varepsilon$.

There are of course many situations where the method doesn't work well. The most obvious is where the Jacobian matrix $\mathbf{f}'$ is singular, in which case the method is not even well defined. The Jacobian may also be almost singular, which leads to an ill-conditioned system (3.2), i.e., a small perturbation in the given data causes a big perturbation in the computed solution. And even if the system is well conditioned, it may be difficult to derive the Jacobian in explicit form. In all of these cases there are various ways to modify the method, which we shall come back to later in the book.

But the fact is that the Newton–Raphson method is the basis for almost all modern methods for solving nonlinear systems of equations. So, here is another case where Newton has had an enormous impact on science.

Carl Gustav Jacob Jacobi
(Carl Gustav Jacob Jacobi,
Nr. 586. Grafiksammlung
Berlin-Brandenburgische
Akademie der
Wissenschaften. All Rights
Reserved)

3.2 Interpolation

Astronomy got an enormous boost during the sixteenth century with the great discoveries by Kopernicus and Kepler and, to a large extent, it was the source of the most significant developments of mathematics for centuries to come. Not only did it lead to the formulation of differential and integral calculus, but also to the development of numerical methods. One obvious challenge was connected to astronomical observations. One example is interpolation as mentioned in Sect. 3.1.2. Even if interpolation doesn't belong to the hard mathematical problems over time, it is a striking fact that most techniques bear the name of some famous mathematician or astrophysicist who developed it.

A representation of the unknown function by polynomials is quite natural. Polynomials were well understood at the time, and easy to evaluate. Given $N + 1$ function values

$$f(x_j), \quad j = 0, 1, \ldots, N,$$

the interpolation problem is to find a polynomial

$$p(x) = a_0 + a_1 x + \ldots + a_N x^N \tag{3.3}$$

such that

$$p(x_j) = f(x_j), \quad j = 0, 1, \ldots, N.$$

The solution to this problem is unique and can be defined by an algebraic system of equations for the $N + 1$ polynomial coefficients a_j:

$$a_0 + a_1 x_0 + \ldots + a_N x_0^N = f(x_0),$$
$$a_0 + a_1 x_1 + \ldots + a_N x_1^N = f(x_1),$$

$$\vdots$$

$$a_0 + a_1 x_N + \ldots + a_N x_N^N = f(x_N).$$

The polynomial value at any point x is then evaluated by (3.3). Accordingly, it may seem that the interpolation problem is no big deal; just solve this system. However, it may not be very easy to solve, in particular for large N. In fact, later it became clear that these systems quickly become ill-conditioned when degree N increases. Therefore, different formulas for evaluating $p(x)$ were derived, and as noted above, many famous mathematicians were involved. Again, Newton was in the frontline, and his interpolation formula is probably the most well known. He wrote the polynomial in the form

$$p(x) = b_0 + b_1(x - x_0) + \ldots + b_N(x - x_0)(x - x_1) \ldots (x - x_{N-1}), \qquad (3.4)$$

where the coefficients b_j have to be computed. Obviously $b_0 = f(x_0)$. It turns out that the remaining coefficients have a very elegant explicit form, as expressed in terms of *divided differences*, which are defined recursively. A divided difference of order j requires $j + 1$ given points

$$x_0, x_1, \ldots, x_j$$

with the corresponding function values

$$f(x_0), f(x_1), \ldots, f(x_j).$$

The points x_j do not have to be ordered in increasing order with increasing j, but it is convenient to have them ordered in that way. The first-order difference using the points x_0, x_1 is

$$f[x_0, x_1] = \frac{f(x_1) - f(x_0)}{x_1 - x_0}, \qquad (3.5)$$

while the differences of order $j \geq 2$ are defined by

$$f[x_0, x_1, \ldots, x_j] = \frac{f[x_1, x_2, \ldots, x_j] - f[x_0, x_1, \ldots, x_{j-1}]}{x_j - x_0}. \qquad (3.6)$$

One can show that the coefficients b_j in (3.4) have the simple form

$$b_0 = f(x_0),$$
$$b_j = f[x_0, x_1, \ldots, x_j], \quad j = 1, 2, \ldots, N,$$

which gives the *Newton interpolation formula*

$$p(x) = f(x_0) + \sum_{j=1}^{N} f[x_0, x_1, \ldots, x_j](x - x_0)(x - x_1) \ldots (x - x_{j-1}). \qquad (3.7)$$

The interpolation polynomial is unique; still many different forms of expressing it were derived following Newton. Interpolation was an important issue, and different forms led to different types of error estimates. Sharp mathematicians, physicists and astronomers were involved, one of them being Joseph Louis Lagrange (1736–1813), who had a quite interesting life story. He had a French father and an Italian mother, and when he started showing interest in mathematics at age 17, he revealed an extraordinary talent. Already at age 19 he got an appointment at the Royal Military Academy of the Theory and Practice of Artillery in Turin. Later he was persuaded to take over the chair after Leonhard Euler in Berlin, where he stayed for the next 20 years. Then he moved to Paris, and when the École Polytechnique opened in 1794, he became a professor there until his death.

Joseph-Louis Lagrange
(©traveler1116/Getty
Images/iStock; 175413925)

Like so many other great minds during this time, Lagrange had such a wide area of interests, not only in mathematics, but also in a number of applied areas such as mechanics and astrophysics. Interpolation techniques may have been a small problem to him, but he came up with the compact and elegant *Lagrange*

interpolation formula

$$p(x) = \sum_{j=0}^{N} f(x_j)\delta_j(x),$$ (3.8)

where

$$\delta_j(x) = \prod_{k=0,\ k\neq j}^{N} \frac{x - x_k}{x_j - x_k}.$$ (3.9)

Observing that

$$\delta_j(x_k) = \begin{cases} 0, & k \neq j \\ 1, & k = j \end{cases},$$

it is clear that the interpolation condition

$$p(x_k) = \sum_{j=0}^{N} \delta_j(x_k)f(x_j) = f(x_k)$$

is satisfied.

In standard interpolation only the function values $f(x_j)$ are used, possibly because nothing more is known about the function. However, the French mathematician Charles Hermite (1822–1901) had the idea that one should use also the derivatives $f'(x_j)$ of the function if these are known. He didn't like functions not being differentiable which he expressed in a letter to Stieltjes:

> I turn with terror and horror from this lamentable scourge of continuous functions with no derivatives (https://en.wikipedia.org/wiki/Charles_Hermite).

Concerning his life story there is one event that is particularly strange. From the viewpoint of today, it is hard to believe that he had to interrupt his studies at the École Polytechnique because he had a deformed right foot. But he got his revenge when coming back there as a professor in 1869. His fame is emphasized by the moon's Hermite crater named after him.

Assume that there are $N + 1$ points x_j with $f(x_j)$ and $f'(x_j)$ known at each of them. The interpolation polynomial $p(x)$ now has degree $2N + 1$, and the coefficients a_j are obtained by the equations

$$p(x_j) = f(x_j), \qquad j = 0, 1, \ldots, N,$$
$$p'(x_j) = f'(x_j), \qquad j = 0, 1, \ldots, N.$$

Charles Hermite Google Books (1901–10). "Charles Hermite". The American Mathematical Monthly VIII: p. 213. Springfield, Missouri: Mathematical Association of America. (Wikimedia*)

As in the standard interpolation case, this system of algebraic equations for a_j may become ill conditioned, and there are different ways of expressing it differently. Indeed, the Newton interpolation formula (3.7) can be used in this case as well. The idea is to count the interpolation points x_j twice and introduce new points

$$y_k, \qquad k = 0, 1, \ldots, 2N + 2$$

such that

$$y_{2j} = y_{2j+1} = x_j, \qquad j = 0, 1, \ldots, N.$$

The divided differences must now be defined for coinciding points $y_{2j} = y_{2j+1}$, and this is done by using the limit

$$f[y_k, y_{k+1}] = \lim_{y_{k+1} \to y_k} \frac{f(y_{k+1}) - f(y_k)}{y_{k+1} - y_k} = f'(y_k).$$

The Newton interpolation formula (3.7) can then be generalized to

$$p(x) = f(x_0) + \sum_{k=1}^{2N+1} f[y_0, y_1, \ldots, y_k](x - y_0)(x - y_1) \ldots (x - y_{k-1}).$$

We demonstrate this with a simple example. Assume that the function $f(x) = x^3$ is represented by the four values:

$$f(0) = 0, \quad f'(0) = 0,$$
$$f(1) = 1, \quad f'(1) = 3.$$

A third degree polynomial $p(x)$ is uniquely defined by four given values; therefore the interpolation formula should give $p(x) \equiv x^3$. The interpolating polynomial is given by

$$p(x) = f(y_0) + f[y_0, y_1](x - y_0) + f[y_0, y_1, y_2](x - y_0)(x - y_1)$$
$$+ f[y_0, y_1, y_2, y_3](x - y_0)(x - y_1)(x - y_2),$$

where

$$f[y_0, y_1] = f'(0) = 0,$$
$$f[y_0, y_1, y_2] = f[y_1, y_2] - f[y_0, y_1] = 1 - f'(0) = 1,$$
$$f[y_1, y_2, y_3] = f[y_2, y_3] - f[y_1, y_2] = f'(1) - f(1) + f(0) = 2,$$
$$f[y_0, y_1, y_2, y_3] = f[y_1, y_2, y_3] - f[y_0, y_1, y_2] = 1.$$

Accordingly,

$$p(x) = 0 + 0 + x^2 + x^2(x - 1) = x^3$$

as expected.

No matter how robust and reliable the method is for computing the coefficients of the polynomials, a fundamental problem arose when going to high degree polynomials using many data points. Certain nonphysical oscillations seemed to always destroy the solution for high enough degree N, and there was no convergence to the precise solution as $N \to \infty$. This was surprising to many, in particular after 1885 when Weierstrass published the Weierstrass theorem, see [160]. It says that for a given continuous function $f(x)$, there is a sequence of polynomials $\{P_N(x)\}$ with uniform convergence on a finite interval $[a, b]$:

$$\lim_{N \to \infty} \left(\max_{a \leq x \leq b} |f(x) - P_N(x)| \right) = 0.$$

A sequence of polynomials defined by an increasing number of equally distributed interpolation points seems to be a convenient choice, but it is not.

Carl Runge (1856–1927) at the University of Hanover explained the source of the trouble in his article [143] published in 1901. He used the quite smooth "Runge function"

$$f(x) = \frac{1}{1 + 25x^2}, \qquad -1 \leq x \leq 1$$

to demonstrate why the interpolating polynomial shows severe oscillations which become worse with increasing N, see Fig. 3.1.

A remedy for this difficulty is to locate the interpolating points differently. This leads naturally to orthogonal polynomials, and we shall come back to them

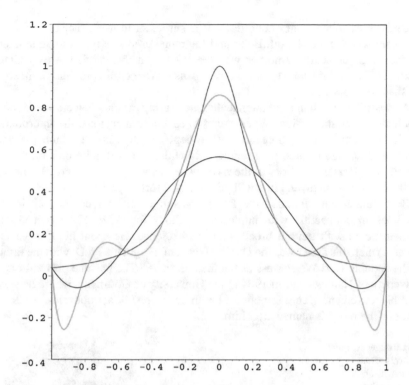

Fig. 3.1 The Runge phenomenon. $f(x)$ (red), $P_5(x)$ (blue), $P_9(x)$ (green)   (Work by Runge. Wikimedia*)

in Sects. 3.3 and 3.6 when discussing numerical integration and series expansion. However, this cure cannot be used if the location of the data points is determined by measurements or observations. A better way is to relax the requirement of polynomial representation on the whole interval. Indeed this was the idea of R. Schoenberg when he introduced piecewise polynomials in the form of splines in 1946. We will come back to this in Sect. 5.5.1.

As an alternative to polynomials, one can of course choose other types of basis functions when interpolating, and trigonometric functions is the classic alternative. In this case the most general form of the interpolating function is

$$p(x) = \sum_{j=0}^{N} a_j \cos(jx) + \sum_{j=1}^{N} b_j \sin(jx),$$

with the coefficients a_j, b_j to be determined. Trigonometric interpolation is of course best suited to functions that are periodic, even if it is well defined for any set of data points. On a uniform grid, this leads to the discrete Fourier transform, and its fast implementation FFT, which will be further discussed in Sect. 3.6.

Under special symmetry conditions, it is sufficient to use either the cos-series or the sin-series. Indeed, both Euler and Lagrange used this type of trigonometric polynomials very early. Another was the French mathematician Alexis Claude Clairaut (1713–1765), who used the cos-expansion for certain problems concerning the shape of the earth.

Almost all work on interpolation techniques during the classical era was done for problems in one dimension. When going to several dimensions, the interpolation problem becomes much more difficult, except for the case in which the given function values are located on a regular grid with each grid point on straight lines in every coordinate direction. In the more general case, piecewise polynomials are much more convenient, and this will be discussed further in Sect. 5.5.1.

Karl Runge, who founded the Runge phenomenon, will come back later in this book in connection with numerical methods for ODE. He was a German mathematician with quite a broad interest in science. He spent his first years in Havana, where his father was the Danish(!) consul. He got his Ph.D. with the famous mathematician Karl Weierstrass as his mentor. In 1886 he became professor at the University of Hanover and in 1904 at the University of Göttingen, where he stayed until his retirement. Being famous and with an interest in astrophysics, the Runge crater on the moon is named after him.

Karl Runge   (Source: Archives of the Mathematisches Forschungsinstitut Oberwolfach)

3.3   Integrals

With the introduction of differential and integral calculus came the problem of computing the value of definite integrals

$$I = \int_a^b f(x)\,dx.$$

Only for special functions $f(x)$ is it possible to find a primitive function $F(x)$ with $dF/dx = f$ by analytical means, such that the formula $I = F(b) - F(a)$ can be applied. Sometimes the explicit form of $f(x)$ is not even known, but its values are known at certain discrete points x_j in the interval $[a, b]$. Therefore it is necessary to construct numerical methods for finding accurate values of I. This is sometimes called *quadrature* from the special technique of finding a square with the same area as that of a given geometrical domain.

As for so many other problems, Newton found ways for the numerical computation of integrals. He had already constructed interpolation formulas by representing a function by polynomials, see Sect. 3.2. It is very natural to use polynomials when approximating $f(x)$ in the integral, since the primitive function of a polynomial is another polynomial that is easily found. In [123] published in 1711, Newton presents his approximation using four points a, b, c, d as

$$\int_a^d f(x)dx = \frac{3(d-a)}{8}\big(f(a) + 3f(b) + 3f(c) + f(d)\big),$$

but he doesn't provide any derivation for this. However, this is the formula obtained when approximating f by a third degree interpolation polynomial.

Roger Cotes (1682–1716) at Cambridge had already worked on the same problem and presented his first results in a public lecture in 1707. When he later read about Newton's result he generalized his own results covering the cases from 3 to 11 points in the integration interval, but it was not published until 1722, see *Harmonia mensurarum*. It may be interesting to see how results were presented at this time; there was a lot of explanatory writing and no notation for a sequence of numbers. He calls his results a set of Rules, and he describes them as follows ([20], translation from Latin):

> The first produces the area lying between the extremes of three equidistant ordinates, the second the area between the extremes of four equidistant ordinates, and Newton gave this, by the third we find the area between five given [ordinates] in the same way, etc. In all these rules, A is used for the sum of the extreme ordinates, that is to say the sum of the first and the last, B for the sum of the [ordinates] closest to the extremes, that is to say the sum of the second and the next to last, C for the sum of the next ones, that is to say the sum of the third and the antepenultimate [ordinates] and so on in the order of the letters D,E,F, etc. until we arrive at the middle ordinate, if there are an odd number of them, or at the sum of the middle ones, if there are an even number of them. Finally, for the interval between the extreme ordinates, that is the base of the required area, we use the letter R.

He then writes down the nine different "Rules", which for the last case with eleven points is

$$\frac{16067\,A + 106300\,B - 48525\,C + 272400\,D - 260550\,E + 427368\,F}{598752}R.$$

(3.10)

Considering the calculations behind this formula, it is easy to understand that he didn't go any further.

Numerical integration based on this principle is called the *Newton–Cotes formulas*. Given $N + 1$ equally spaced points

$$\left(x_j, f(x_j)\right), \qquad j = 0, 1, \ldots, N,$$

the approximate integral is obtained by integrating the N-degree polynomial $p_N(x)$ passing through these points. The polynomial is unique, but there are many ways to evaluate it, and the formulas are commonly defined in terms of Lagrange polynomials. Referring back to Sect. 3.2, the interpolation polynomial can be written in the form (3.9), (3.8), which leads to the integration formula

$$I_N = \sum_{j=0}^{N} w_j f(x_j), \qquad w_j = \int_a^b \delta_j(x)$$

with $\delta_j(x)$ defined in (3.9). The weights w_j are obtained by simple integral calculus. The resulting first two formulas are the *trapezoidal method*

$$I_1 = \frac{b - a}{2}\left(f(x_0) + f(x_1)\right)$$

and *Simpson's rule*

$$I_2 = \frac{b - a}{6}\left(f(x_0) + 4f(x_1) + f(x_2)\right).$$

The name of the second refers to the British mathematician Thomas Simpson (1710–1761) who used it probably without knowing about the general Newton–Cotes formulas. Actually, it seems like this particular formula was used already by Kepler in connection with his work in astronomy.

In general the interval $[a, b]$ is too large for these coarse approximations. The polynomial corresponding to the Newton–Cotes formula (3.10) is of degree 10, which seems to be quite high. However, if the function f varies rapidly, it is still not sufficient for resolving the oscillations. As an example, consider the integral

$$\int_0^{\pi/2} k \sin(kx)\, dx,$$

which has the value 1 if $k = 1, 5, 9, \ldots$. The following table shows the error with the method (3.10):

k	$Int(k)$
1	$1.0 \cdot 10^{-13}$
5	$9.7 \cdot 10^{-5}$
9	0.11
13	4.8

If the degree of the polynomial is not sufficient to resolve the highly oscillatory function above, we know already from Sect. 3.2 that the use of higher order polynomials is the wrong way to go when looking for better accuracy, since the Runge phenomenon shows up quite early for certain functions. Instead, the interval of integration is divided into many subintervals with the integration formula applied to each of them. For example, Simpson's rule is first used on $[x_0, x_2]$, then on $[x_2, x_4]$, etc. If the stepsize is h, it can be shown that the aggregated error for the approximation $\tilde{I}$ is

$$|\tilde{I} - I| \le Ch^4$$

for Simpson's rule. At first glance this is surprising, since the method is based on interpolation polynomials of degree 2. These normally provides an error of order h^3, but the symmetry of the integration formula raises the accuracy one step.

If the location of data points is not restricted, there is another way to improve performance, still using polynomials for interpolation. In 1814, one of the greatest mathematicians of all time, Carl Friedrich Gauss (1777–1855), published the article [63], in which he describes how orthogonal polynomials are used for integration. As an example, we take the interval $[-1, 1]$ and the computation of $\int_{-1}^{1} f(x) \, dx$. In Sect. 3.4 the Legendre orthogonal polynomials $p_N(x)$ are defined. Gauss made the beautiful observation that if the nodes x_j are chosen as the N roots of $p_N(x)$, then the integration formula

$$I_N = \sum_{j=0}^{N} w_j f(x_j), \qquad w_j = \frac{2}{(1 - x_j^2) p_N'(x_j)^2}$$

is exact for all polynomials

$$f(x) = p_k(x), \qquad k = 0, 1, \ldots, 2N - 1.$$

This is remarkable. We are working with polynomials of degree N, but achieving accuracy that corresponds to degree $2N - 1$.

The construction can be generalized to any integral $\int_a^b u(x) f(x) \, dx$ by choosing the proper orthogonal polynomials corresponding to the weight function $w(x)$. A robust and efficient way of computing the weights w_j in the summation formula was presented by Golub and Welsch (1969), see [75].

Numerical methods for integrals in several dimensions d can be constructed by using the same principle as above. However, just as for interpolation, the generalization is not trivial, in particular when the domain of integration is not a rectangle. By choosing other discretizations, for example based on subdivision into triangles, approximations using piecewise polynomials are usually more efficient. Furthermore, Monte Carlo methods, as described in Sect. 5.14.1, may be the best alternative, particularly when d is large.

Finally we note that formulas directly connected to numerical integration were already derived by the Egyptians around 1890 BC! These results are collected in the Golenischev Mathematical Papyrus written on a papyrus roll now kept at the Pushkin State Museum of Fine Arts in Moscow.

3.4   The Least Squares Method

Astronomy continued to be the driving force behind computational mathematics for centuries. Not only was it of prime interest from a scientific point of view in order to understand the universe, but it also had practical implications. Sea travel was necessary when exploring the globe and when claiming new colonies. But how could one navigate accurately when travelling the oceans without any land in sight? The stars were the only help.

But astronomy by itself was big science. The big planets had been discovered, and Newton's theory was the basis for determining their orbits. However, there were small perturbations of theoretical calculations that indicated the existence of very small planets called planetoids or asteroids depending on their size. At the end of the eighteenth century these perturbations showed that there must be a large body between Mars and Jupiter. The Italian astronomer Giuseppe Piazzi (1746–1826) found it in January 1801 using his telescope in Palermo, and it was given the name Ceres. He made many observations during a little over a month, but then he lost it because it went too close to the sun. Many astronomers set out to calculate the orbit such that it could be found again. Gauss became interested in the problem, and he made the only assumption that the orbit was an ellipse. There were 6 unknown parameters representing the position of the ellipse, the eccentricity and the position of Ceres at a certain time. The computation of a unique solution required 6 observations, and many more were available. We have earlier described interpolation methods, in which polynomials are used as the basis functions. These polynomials are usually not the true functions; they are used simply because they are convenient to handle. If the formulas are used for extrapolation, one is evaluating the interpolating function outside the interval where the true function is known, then the precision goes down dramatically. In the Ceres case, the parameters to be determined describe the orbit exactly. However, the accuracy of the measurements was not very good; furthermore they were all obtained quite close to each other. This meant that the resulting calculated orbit could be way off the true one. Instead Gauss used more observations and introduced the least squares method, and we shall now describe it.

The method is quite general, but in the case of Gauss, it was developed for the special case with an overdetermined linear system of algebraic equations

$$Ax = b.$$

Here $\mathbf{x}$ is a column vector with p elements to be determined, and A is a rectangular matrix with N rows and p columns, $N > p$. In the Ceres problem $p = 6$, $N = 11$ in the final calculation. In general there is of course no solution to this problem, so we define a least squares solution by computing

$$\min_{\mathbf{x}} \sum_{i=1}^{N} \left(\sum_{j=1}^{p} a_{ij} x_j - b_i \right)^2 .$$

This minimum is obtained by requiring that the partial derivatives with respect to x_j are zero:

$$\sum_{i=1}^{N} \left(\sum_{j=1}^{p} a_{ij} x_j - b_i \right) a_{ij} = 0, \qquad k = 1, 2, \ldots, p,$$

which can be written in the simple form

$$A^T A \mathbf{x} = A^T \mathbf{b}.$$

This system is called the normal equations and it has a unique solution which solves our least squares problem. The solution to this and other linear systems was a challenge to Gauss. In Sect. 3.5 we shall show how he solved it, and how others followed up with refined methods.

Gauss worked with Piazzi on the concrete problem. The calculations were heavy, but they soon came up with the solution. Many others came up with their own predictions as well, and several of them were published by Franz Xaver von Zach in September 1801. Gauss's prediction was quite different from the others, but he was right. In December, the same year Ceres was rediscovered and the position was almost exactly where Gauss had predicted.

At this time he was only 24 years old, but this achievement made him world famous. Somewhat surprising, he didn't publish his method until 1809 in [62], which contained a quite general theoretical presentation of celestial mechanics. It was first writing in German, which was then translated into Latin in order to get a wider audience. In the English translation, the title is "Theory of motion of the celestial bodies moving in conic sections around the Sun".

There is actually a special story concerning who should get credit for the least squares method. During this time Adrien-Marie Legendre (1752–1833) worked independently of Gauss on the problem of computing orbits of the celestial bodies.

The first publication [113] containing an explicit presentation of the least squares method was published in 1805 by Legendre. The book was in French, in translation the title is "New methods for the determination of the orbits of comets". Gauss claimed that he had already developed the method without publishing it. Furthermore, the 1809 book by Gauss contained new results concerning statistical considerations. He proved that under the assumption that the errors in the observations have a normal distribution (also called the Gauss distribution) independent of each other, then the least squares method provides the most probable estimate of the solution.

Gauss and Legendre were both primarily interested in the particular problem of over-determined linear systems of equations. However, the least square method is more general than that, and it can be applied to any set of data representing a function that is known only at certain discrete points. Polynomials are common when approximating functions, but not well suited for interpolation at many points, since the degree goes up with an increasing number of data, and as we have seen above, this leads to ill-conditioned problems with the possibility of nonphysical oscillations in the result. So how could one keep the degree of the polynomials low and still use all the information available? The answer is the least squares method, which allows for errors in the approximation, and we shall demonstrate it for a function of one variable. The procedure is often called *curve fitting*.

Assume that there is a function $f(x)$ with known values $f_j = f(x_j)$, $j = 1, 2, \ldots, N$ at a uniformly distributed sequence of points $\{x_j\}$ with a distance h between them. The scalar product between two sets of function values $\{f_j\}$ and $\{g_j\}$ is defined by the sum

$$(f, g) = \sum_{j=1}^{N} f_j g_j h,$$

and the *norm* by

$$\|f\| = \sqrt{(f, f)}.$$

With basis functions $\phi_j(x)$, the least square approximation is

$$g(x) = \sum_{j=1}^{p} a_j \phi_j(x),$$

where $p < N$, and we seek

$$\min_{a_j} \|f - g\|^2.$$

It is easy to show that the minimum is obtained when the error is orthogonal to each basis function, i.e.,

$$(f - g, \phi_k) = 0, \qquad k = 1, 2, \ldots, p.$$

This gives the system

$$\sum_{j=0}^{p} (\phi_j, \phi_k) a_j = (f, \phi_k), \qquad k = 1, 2, \ldots, p. \tag{3.11}$$

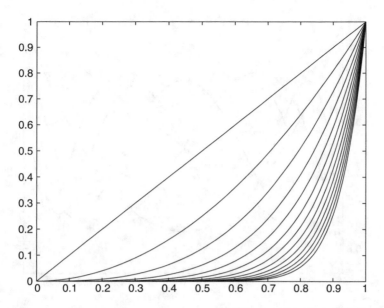

Fig. 3.2 Polynomials x^p, $p = 1, 2, \ldots, 12$

As noted above, one should avoid polynomials with basis functions

$$\phi_j(x) = x^{j-1}, \qquad j = 1, 2, \ldots, p$$

if the degree p is not very small. The reason is that they become almost linearly dependent, which leads to an ill-conditioned system (3.11). Indeed it is easy to see that these basis functions are no good by taking a look at the graph representing them. Figure 3.2 shows the functions x, $x^2, \ldots, x^{12}$.

Increasing the degree of the polynomials is characterized by a smaller value for any given x except for the endpoints. There is not much difference between the higher order ones, and clearly they are not well suited to represent other functions no matter how they are combined. The cure is to work with another basis, and the best result is obtained with orthogonal polynomials. Here we define orthogonality between two functions $f(x)$ and $g(x)$ on the interval $[-1, 1]$ by using integrals as

$$\int_{-1}^{1} f(x)g(x)\,dx = 0.$$

Actually, Legendre himself constructed a special type of orthogonal polynomials later named after himself. They are defined on the interval $[-1, 1]$ and can be

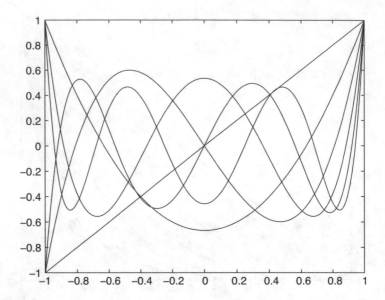

Fig. 3.3 Legendre polynomials of degree $p \leq 6$

generated recursively as follows for degree p:

$$\phi_0(x) = 1,$$
$$\phi_1(x) = x, \tag{3.12}$$
$$j\phi_j(x) = (2j - 1)x\phi_{j-1}(x) - (j - 1)\phi_{j-2}(x), \qquad j = 2, 3, \ldots, p.$$

Figure 3.3 shows these polynomials up to degree 6, where an increasing degree can be identified by more fluctuations in the interval.

Each polynomial is orthogonal to every other polynomial, and they are much better as basis functions when using the least squares method. We come back again to this in Sect. 4.4.1.

Finally a few remarks about Gauss. He was a true "Wunderkind", i.e., a wonder-child, who already as a child had developed mathematical skills that were far above any normal level. One of his early accomplishments was the solution of a quite unusual problem, namely finding out the date of his birth. His mother was illiterate, but remembered that he was born on a Wednesday eight days before the Feast of the Ascension which occurs 39 days before Easter. At this time is was not trivial to find out when Easter occurs in the past and in the future, but Gauss solved it and determined his birthday as April 30.

Carl Friedrich Gauss
(Source: Universität
Göttingen, Sammlung
Sternwarte. Oberwolfach
Photo Collection)

At age 15 he had already entered Braunschweig University of Technology (called Collegium Carolinum at the time) and at 18 he transferred to the University of Göttingen. He had a very wide mathematical spectrum of interests, and proved very early fundamental and deep theorems. At age 19 he solved a geometry problem that had been considered for a long time. He proved that a polygon with equal sides can be constructed if and only if the number of sides is the product of distinct Fermat primes and a power of two. The discovery made him so proud and happy that he requested that a regular heptadecagon (17 sides) should be engraved on his tombstone. However, his request was refused, because the technique didn't allow for a clear distinction between a heptadecagon and a circle.

At age 22 he proved one of the most well-known fundamental theorems called the *fundamental theorem of algebra*. It says that every nonconstant polynomial of one complex variable and with complex coefficients has at least one complex root. (From this it follows easily that an n-degree polynomial has exactly n roots.) Actually he used an assumption that was not proved, and he later came up with a new proof.

His unusual breadth of research consists of geometry, topology, algebra, number theory, probability theory on the pure side, and various numerical methods on the more applied side. Furthermore, he applied his theories and methods to several areas in science and, as usual at the time, astronomy was the dominating topic. He also built various physical instruments, among others an instrument for measuring the magnetic field around the earth.

There are many examples of certain achievements by other mathematicians where Gauss claimed that he had already worked out the same proof or method. The least squares method described above is such an example. It may seem like an unfair trick to make such a claim afterwards, but his assertions were probably true in most cases. He seems to have avoided publishing a certain result if he didn't consider it complete even if it was fundamentally a new discovery.

His work on the orbit of Ceres became widely known among astronomers society. Mathematics as a field of research was not well established at the time, and Gauss himself was not sure about its future. Therefore in 1807 he accepted an appointment as Professor of Astronomy, and not only that, he was also appointed Director of the astronomical observatory in Göttingen.

Finally, one can note that Gauss had a quite unusual attitude regarding the career of his sons (according to a letter 1898 by his grandson Charles Henry Gauss). He didn't want them to become mathematicians since he judged them to be less talented than their father, and he feared that they would lower the family name.

3.5 Gauss and Linear Systems of Equations

As described in the previous section, Gauss used the least squares method that led to the normal equations which had six unknowns in his first example. Indeed he used various methods for solving them, and iteration was a basic ingredient. However, direct elimination methods without iteration for general $N \times N$ systems are usually referred to as Gaussian elimination, even if the first known presentation of this technique is actually very old. The Chinese mathematician Liu Hui (225–295) described it in his comments on the book *The Nine Chapters on the Mathematical Art*. Gauss began discussing the method in [62] and followed up with certain remarks in later articles.

The basic method is being taught already in high school for small systems. The idea is simply to use the first equation to eliminate the first variable from the remaining $N - 1$ equations, and then use the modified second equation to eliminate the second variable from the remaining $N - 2$ equations, and so on. The new system can then be solved by back-substitution beginning with the last equation and then using the computed values of x_N, x_{N-1}, ... working upwards in the system.

The procedure can be seen as transforming the original system

$$A\mathbf{x} = \mathbf{b} \tag{3.13}$$

to another system

$$QA\mathbf{x} = Q\mathbf{b},$$

where $U = QA$ is an upper triangular matrix. The first element of the modified right-hand side contains only the first element b_1 of the original one, the second element contains only b_1 and b_2, etc. This means that Q is a lower triangular matrix, and as a consequence, $L = Q^{-1}$ is lower triangular as well. Since $A = LU$, Gaussian elimination is actually a factorization of the original system, such that

$$LU\mathbf{x} = \mathbf{b},$$

where L and U are lower- and upper-triangular respectively. This procedure goes under the name LU decomposition.

A straightforward implementation of Gaussian elimination may not work for obvious reasons. At a certain stage k of the elimination procedure, the variable x_k may not be present in equation no. k. In such a case, we can reorder either the equations or the columns of the matrix. This is called row pivoting in the first case and column pivoting in the second case. If the diagonal element is not exactly zero but small in magnitude, pivoting is necessary as well in order to avoid the bad influence of rounding errors. Therefore, any standard system solver uses pivoting, usually combined pivoting, where at each stage the largest element in the lower submatrix is found followed by row and column switches, such that the maximal element becomes the new diagonal element.

The Gaussian elimination procedure requires approximately $2N^3/3$ arithmetic operations, but in the case of a symmetric matrix A, one can do better. This was found by André-Louis Cholesky (1875–1918), but his accomplishment has a somewhat sad history. His parents were Polish (spelling their name as Cholewski) emigrated to France. He attended the École Polytechnique and became interested in geodesy, in particular making accurate maps, which led to large linear symmetric systems of equations. But he was an engineering officer in the French army and served in the first world war, where he was killed. A complete biography is given in the book [14].

André-Louis Cholesky
(Springer International
Publishing Switzerland)

The Cholesky recursive algorithm is as follows. Assume that at stage j, the matrix A_j has the form

$$A_j = \begin{bmatrix} I_{j-1} & 0 & 0 \\ 0 & a_{jj} & \mathbf{c}_j^T \\ 0 & \mathbf{c}_j & C_j \end{bmatrix},$$

where I_{j-1} is the $(j-1) \times (j-1)$ identity matrix. With $A_1 = A$, the vector $\mathbf{c}_1$ and the matrix C_1 are well defined. With

$$L_j = \begin{bmatrix} I_{j-1} & 0 & 0 \\ 0 & \sqrt{a_{jj}} & 0 \\ 0 & \mathbf{c}_j/\sqrt{a_{jj}} & I_{N-j} \end{bmatrix}$$

and

$$A_{j+1} = \begin{bmatrix} I_{j-1} & 0 & 0 \\ 0 & 1 & 0 \\ 0 & 0 & C_j - \mathbf{c}_j\mathbf{c}_j^T \end{bmatrix},$$

the identity

$$A_j = L_j A_{j+1} L_j^T$$

holds. Since $A_{N+1} = I$, we obtain the lower triangular matrix L in $A = LL^T$ as

$$L = L_1 L_2 \dots L_N.$$

For Hermitian complex matrices, the same formula holds with the transpose C^T of a matrix C substituted by the conjugate transpose C^* everywhere. One can show that computing Cholesky factorization can be done with $N^3/3$ arithmetic operations, which is twice as fast as standard Gaussian elimination.

Many applications today lead to very large systems with millions of unknowns, and the operation count $\mathcal{O}(N^3)$ makes direct elimination methods unrealistic even with high performance computers. In the nineteenth century one usually worked on systems of more modest size, but since no computers were available, hand calculation became quite tedious anyway. Therefore Gauss used various kinds of iteration procedures depending on the particular type of system. He may have computed an approximation to a certain subset of unknowns and then used some procedure to compute the remaining unknowns in terms of the first subset. This type of iteration is then continued in the same manner.

A systematic way of using old recently computed unknowns is the following. The matrix A in (3.13) is decomposed as

$$A = L + U,$$

where L is the lower triangular part of A including the diagonal. The system is then rewritten as

$$L\mathbf{x} = \mathbf{b} - U\mathbf{x}.$$

With an initial guess $\mathbf{x}^{(0)}$, the first equation is now solved for the first element $x_1^{(1)}$ which is substituted for the first element of $\mathbf{x}$ in the second equation. We can then solve for the second element $x_2^{(1)}$. This procedure is continued such that at each stage the latest available element in $\mathbf{x}$ is used. The algorithm can be written in compact form as

$$\mathbf{x}^{(k+1)} = L^{-1}(\mathbf{b} - U\mathbf{x}^{(k)}), \qquad k = 0, 1, \ldots.$$

Each iteration has a simple structure, and requires only $\mathcal{O}(N^2)$ arithmetic operations. If the convergence rate is good, the method is much faster than direct elimination.

The method was communicated by Gauss in a letter to one of his students, but it was never published. Later Phillip Ludwig von Seidel (1821–1896) picked up the idea and published it much later.

The Germans were really in the forefront when it came to methods for solving systems of algebraic equations. We have already mentioned Carl Jacobi in Sect. 3.1.2 when discussing the solution of nonlinear systems. He generalized the Newton iterative method for nonlinear scalar equations to nonlinear systems of equations, which led to the solution of linear systems in each iteration step. He then used what is now called the *Jacobi method*, which is actually a simpler version of the Gauss-Seidel method, and it was published in 1845, see [94]. The matrix A is now partitioned as

$$A = D + R,$$

where D contains only the diagonal of A. The algorithm is

$$\mathbf{x}^{(k+1)} = D^{-1}(\mathbf{b} - R\mathbf{x}^{(k)}), \qquad k = 0, 1, \ldots.$$

For both methods there is the necessary requirement that the diagonal of A be nonzero, but convergence requires more. One sufficient condition is that A be *diagonally dominant*, i.e.,

$$|a_{ii}| > \sum_{j \neq i} |a_{ij}|, \qquad i = 1, 2, \ldots, N.$$

For solving linear systems of equations, we have again an example where great mathematicians came up with new ideas that are still the fundamental basis in modern times for effective computational methods. In Sect. 4.4.2 we will further discuss such methods.

3.6   Series Expansion

This entire book is about the history of numerical methods, which today are the
only methods for computation that are used in practice. However, one of the most
powerful techniques in applied mathematics is series expansion. These techniques
were developed long before computers, partly in order to understand the properties
of the solutions. These series are in general infinite, but if the convergence properties
are reasonable, they can also be used for computation by truncation after a finite
number of terms. But the most powerful techniques are based on modified versions
of the expansions, which are connected to some kind of discrete representation of
functions. This is the reason why these expansions and transforms have played such
an important role in modern computational methods.

Series expansions were used by Newton and Leibniz, in an attempt to obtain π
with high degree of accuracy, which long after Archimedes was a problem of prime
interest among mathematicians. They both used a technique in which a certain part
of the circumference of a circle was found as a function of the diameter. In the old
days scientific achievements were often documented in letters to colleagues, and
in this case Leibniz wrote to La Roque and gave an expression for a fourth of the
circumference of a circle with diameter 1 as

$$\frac{\pi}{4} = 1 - \frac{1}{3} + \frac{1}{5} - \frac{1}{7} + \frac{1}{9} - \frac{1}{11} + \dots .$$

This letter was sent in 1673 but the formula together with more general results was
not published until 1682, when it occurred with a Latin title in *Acta Eruditorum*.
Newton's result containing a series expansion for $\pi/24$ was published in his
Methods of Fluxions.

More general types of series expansions hold on a whole interval $[a \le x \le b]$.
They are based on a sequence of basis functions $\{\phi_j(x)\}_j$, and a certain function
$f(x)$ is represented as an infinite series

$$f(x) = \sum_j c_j \phi_j(x),$$

where the coefficients c_j are constants that are to be computed. In many applications
the analysis simplifies considerably by dealing with the coefficients c_j instead of the
function $f(x)$. One can see the sequence of coefficients $\{c_j\}$ as a *transformation* or
transform of the function $f(x)$.

In the following we shall discuss the most common types of basis functions and
the properties of the associated expansions.

3.6.1   Taylor Series

The best known series expansion is probably the Taylor series named after the
English mathematician Brook Taylor (1685–1731). Given a function $f(x)$ with k
continuous derivatives in the neighborhood of a point $x = a$, the Taylor polynomial
has the form

$$P_k(x) = f(a) + \frac{df}{dx}(a)(x - a) + \frac{1}{2!}\frac{d^2 f}{dx^2}(a)(x - a)^2 + \ldots + \frac{1}{k!}\frac{d^k f}{dx^k}(a)(x - a)^k.$$

If all derivatives exist, i.e., f is analytic, we can pass to infinity and obtain the Taylor
series

$$T_f(x) = \sum_{j=0}^{\infty} \frac{1}{j!}\frac{d^j f}{dx^j}(a)(x - a)^j.$$

The question of convergence of this series, and to what function it converges, is
somewhat complicated and we don't go further into this here. From a numerical
point of view, the finite expansion is of great interest to analyze, and it is actually
frequently used for analyzing many different numerical methods. One of them is
finite difference methods for differential equations, where $|x - a|$ is small. The
basic Taylor theorem says:

$$f(x) = P_k(x) + R(x, a),$$

where the error $R(x, a)$ tends to zero faster than $|x - a|^k$ as $x \to a$. If f is
differentiable $k + 1$ times, we have with $|x - a| = h$

$$|f(x) - P_k(x)| \le Ch^{k+1},$$

where C is a constant. Looking for an approximation of the derivative, we have

$$\frac{f(a + h) - f(a)}{h} = \frac{df}{dx}(a) + \mathcal{O}(h),$$

which is called a first-order approximation.

Taylor expansion is a central tool that is used again and again for analyzing
numerical methods when it comes to accuracy. It is easy to apply, but one should
keep in mind that the theory requires a certain smoothness in the functions. When a
certain method is said to have accuracy of order k, it is based on the assumption
that the function is sufficiently smooth, while real accuracy may be less due to
nonsmooth solutions.

The history of Taylor expansions is quite old. Even if it is named after Taylor,
it had been used much earlier for special functions. The Indian mathematician

and astronomer Madhava of Sangamagrama (1340–1425) (some uncertainty of the exact years) used series expansion of special functions, among them the trigonometric functions. Probably unaware of this work, the Scottish mathematician and astronomer James Gregory (1638–1675) formulated it for more functions, but the first complete construction for general functions was done by Taylor.

The Scottish mathematician Colin MacLaurin (1698–1746) used Taylor series for analyzing analytical functions, and he was probably the one who gave it its name even if the technique was known before Taylor. He worked using the special case with f varying around $x = 0$, i.e., $a = 0$ above, and for some reason he got his name attached to this case, even if the MacLaurin series is just a special case of the Taylor series.

Finally we note that the Taylor series exists in any number of dimensions. For example, the two-dimensional Taylor polynomial of second order is

$$P_2(x, y) = f + (x - a)\frac{\partial f}{\partial x} + (y - b)\frac{\partial f}{\partial y}$$

$$+ \frac{1}{2}\left((x - a)^2\frac{\partial^2 f}{\partial x^2} + 2(x - a)(y - b)\frac{\partial^2 f}{\partial x \partial y} + (y - b)^2\frac{\partial^2 f}{\partial y^2}\right),$$

where f and all its derivatives are taken at the point $(x = a, \ y = b)$.

3.6.2 Orthogonal Polynomial Expansions

Orthogonality between two functions $f(x)$ and $g(x)$ on an interval (a, b) is in general defined through a scalar product

$$(f, g) = \int_a^b c(x)f(x)g(x)dx,$$

where $c(x)$ is a positive weight function. If $(f, g) = 0$, the two functions are orthogonal. With the norm defined by

$$\|f\|^2 = \int_a^b c(x)|f(x)|^2 dx,$$

they are called orthonormal if $\|f\| = \|g\| = 1$. A whole set of functions $\{\phi_j(x)\}_{j=0}^N$ is orthogonal if

$$(\phi_j, \phi_k) = 0, \qquad j \neq k,$$

and orthonormal if

$$\|\phi_j\| = 1, \qquad j = 0, 1, \ldots, N.$$

By defining different forms of scalar products with different intervals (a, b) and different weight function $c(x)$, one can define various sets of orthogonal functions.

Orthogonality can also be defined for discrete functions just as a generalization of orthogonal vectors. The scalar product is defined for two discrete functions $\{f_j = f(x_j)\}$ and $\{g_j = g(x_j)\}$ by

$$(f, g)_h = \sum_{j=0}^{N} c_j f_j g_j h,$$

where the subscript h indicates the discrete character of the scalar product. Here it is assumed that the points x_j are uniformly distributed with $x_{j+1} = x_j + h$. We shall come back to this later when discussing Fourier series.

There are many sets of orthogonal functions, in particular polynomials, that have played a significant role in applied mathematics. We mention here two types of polynomials that have had a remarkable revival in computational mathematics as we shall see later. The first type is the Legendre polynomial $P_n(x)$ defined by the recursion formula (3.12). These real polynomials are orthogonal under the basic scalar product

$$(f, g) = \int_{-1}^{1} f(x)g(x)dx.$$

Pafnuty Chebyshev
(Source: Russian Academy of
Science)

The others are the Chebyshev polynomials constructed by the Russian mathematician Pafnuty Chebyshev (1821–1894). Also here there is a recursive definition:

$$T_0(x) = 1,$$
$$T_1(x) = x,$$
$$T_{n+1}(x) = 2xT_n(x) - T_{n-1}(x), \qquad n = 1, 2, \ldots.$$

The Chebyshev polynomials are orthogonal under the scalar product

$$(f, g) = \int_{-1}^{1} \frac{1}{\sqrt{1 - x^2}} f(x)g(x)\, dx.$$

The classical orthogonal polynomials show a remarkable mathematical beauty in various ways. One property of the Chebyshev polynomials that is useful for numerical analysis is the relation

$$T_n(\cos\theta) = \cos(n\theta), \qquad n = 0, 1, \ldots,$$

or, equivalently,

$$T_n(x) = \cos(n \arccos x).$$

This relation shows that the polynomials have an oscillating character, and that

$$\max_{|x|\leq 1} |T_n(x)| \leq 1, \qquad n = 0, 1, \ldots.$$

Furthermore, the polynomials are orthogonal under the discrete scalar product

$$(f, g)_h = \sum_{j=0}^{N-1} f(x_j)g(x_j)h,$$

where $h = 2/N$ and

$$x_j = \cos\frac{\pi(j + 1/2)}{N}$$

are the *Gauss–Lobatto* points satisfying

$$T_N(x_j) = 0, \qquad j = 0, 1, \ldots, N - 1.$$

Upon a first look at the discrete scalar product $(f, g)_h$, it may be surprising that all points have equal weight since the continuous scalar product (f, g) has the weight function $1/\sqrt{1 - x^2}$ which is very large near $x = \pm 1$. However, the points

x_j are located very tightly at the ends of the interval, which is the way to achieve the stronger weight there.

3.6.3 Fourier Series

The most famous and most frequently used transform is certainly the Fourier transform invented by the French mathematician Joseph Fourier (1768–1830). In this case the basis functions are the complex functions $\phi_\omega(x) = e^{i\omega x}$, and we have with the commonly used notation $c_\omega \to \hat{f}_\omega$

$$f(x) = \sum_{\omega=-\infty}^{\infty} \hat{f}_\omega e^{i\omega x}. \tag{3.14}$$

The formula for the coefficients is the simple inverse Fourier transform

$$\hat{f}_\omega = \frac{1}{2\pi} \int_0^{2\pi} f(x) e^{-i\omega x} \, dx.$$

Since $f(x)$ is represented as a sum, the representation (3.14) is called a Fourier series, and obviously it holds only for periodic functions satisfying $f(x + 2\pi) = f(x)$ for all x. The period 2π can of course be changed to an arbitrary length by a simple transformation.

There are a few names from the classical era that occur more frequently in mathematical work than others. Joseph Fourier is certainly one of them, and he has a very interesting history. He was born in 1768, and was educated by the Benedictine Order. He was active in the French Revolution, and during the "Terror" following the revolution he was put in prison for a short period. However, in 1795 he got an appointment at the École Normale in Paris, which was closely related to the revolution, and later he went to the École Polytechnique in Palaiseau.

In 1798 he joined Napoleon Bonaparte as a scientific advisor on his Egyptian expedition. There he became secretary at the Cairo Institute created by Napoleon for the purpose of strengthening French influence in the Middle East. After the French defeat against the British, Fourier returned to France and became the governor ("prefect") of the Department of Isère in Grenoble. This sounds a bit strange for a mathematician, but he was highly regarded by Napoleon who wrote: "...*the Prefect of the Department of Isère having recently died, I would like to express my confidence in citizen Fourier by appointing him to this place.*" (MacTutor History of Mathematics archive.) In 1816 he moved to England, but returned 6 years later to become the Permanent Secretary of the French Academy of Sciences.

Fourier is certainly best remembered for the Fourier transform, but he made major contributions in physics, primarily with his development of models and analysis of heat conduction, see [53].

Finally we note the striking mentor/student connection between three famous mathematicians; Lagrange was Fourier's mentor, Dirichlet (see Sect. 4.3.2) was his student. France had an extraordinarily strong period in mathematics at this time.

**Jean-Baptiste Joseph
Fourier**   (©Science Photo
Library; 11828512)

The Fourier series represents a spectral analysis of the function $f(x)$, seen as composed of different waves. Each component

$$e^{i\omega x} = \cos(\omega x) + i \sin(\omega x)$$

represents a certain part of the function with wave number (or frequency) ω, the corresponding coefficient $\hat{f}_\omega$ represents the strength of this particular part. We are dealing with complex numbers here. However, if $f(x)$ is real, we have $\hat{f}_{-\omega} = \overline{\hat{f}_\omega}$, and it follows that with $\hat{f}_\omega = a + bi$

$$\hat{f}_\omega e^{i\omega x} + \hat{f}_{-\omega} e^{-i\omega x} = (a + bi)(\cos \omega x + i \sin \omega x) + (a - bi)\big(\cos(-\omega x) + i \sin(-\omega x)\big)$$

$$= 2a \cos \omega x - 2b \sin \omega x.$$

Therefore a real function $f(x)$ can be written in the form

$$f(x) = \sum_{\omega=0}^{\infty} c_\omega \cos \omega x + \sum_{\omega=1}^{\infty} s_\omega \sin \omega x \,,$$

where

$$c_\omega = 2 \, Re \, \hat{f}_\omega \,, \qquad \omega = 0, 1, \ldots,$$

$$s_\omega = -2 \, Im \, \hat{f}_\omega \,, \qquad \omega = 0, 1, \ldots.$$

However, Fourier analysis has had an impact on mathematical and numerical analysis that goes way beyond a straightforward spectral analysis, particularly when dealing with differential equations. Why is that?

The fundamental reason is that differentiation is so simple when working in Fourier space. We have

$$g(x) = \frac{df}{dx}(x) = \frac{d}{dx} \sum_{\omega=-\infty}^{\infty} \hat{f}_\omega e^{i\omega x} = \sum_{\omega=-\infty}^{\infty} i\omega \hat{f}_\omega e^{i\omega x},$$

i.e., the differential operator d/dx is substituted by simple multiplication of the Fourier coefficients:

$$g(x) = \frac{df}{dx}(x),$$

$$\hat{g}_\omega = i\omega \hat{f}_\omega.$$

As an example of Fourier analysis, we choose the simplest form of heat conduction in a one-dimensional space governed by the partial differential equation

$$\frac{\partial u}{\partial t} = \frac{\partial^2 u}{\partial x^2},$$

where $u = u(x, t)$ is the temperature with the initial distribution $u(x, 0)$. In this case we use the Fourier expansion in the space

$$u(x, t) = \sum_{\omega=-\infty}^{\infty} \hat{u}_\omega(t) e^{i\omega x},$$

which results in a system of ordinary differential equations for the time dependent Fourier coefficients

$$\frac{d\hat{u}_\omega}{dt}(t) = -\omega^2 \hat{u}_\omega(t), \qquad \omega = 0, \pm 1, \pm 2, \dots.$$

Each coefficient has the simple form

$$\hat{u}_\omega(t) = e^{-\omega^2 t} \hat{u}_\omega(0), \tag{3.15}$$

and by using the inverse Fourier transform, the solution is obtained as an infinite series in physical space.

The form of the Fourier coefficients in (3.15) demonstrates the typical property of heat conduction problems. There is a damping of the amplitude with time for each nonzero wave number, and the damping is stronger for higher wave numbers.

Initial temperature distributions that are nonsmooth, for example containing discontinuities, give rise to large amplitudes for large wave numbers. However, because of the strong damping properties, the temperature will quickly become smooth as a function of x. The Fourier technique not only allows for an explicit solution of the problem, it also illuminates essential properties of the solution.

Orthogonality is an important mathematical concept that was used early in applied mathematics. It has a geometrical meaning from the beginning, but later got a much more general meaning. We discussed orthogonal polynomials in Sect. 3.6.2 representing real functions, but when it comes to Fourier series, we consider complex basis functions. By periodicity we limit the function to the interval $[0, 2\pi]$ and define the scalar product between two complex functions f and g by

$$(f, g) = \int_0^{2\pi} \overline{f}(x)g(x)\, dx.$$

Orthogonality is defined as in the real case, and a set of complex functions $\{\phi_\omega(x)\}_\omega$ is orthogonal if

$$(\phi_\omega, \phi_v) = 0, \qquad \omega \neq v,$$

and orthonormal if

$$(\phi_\omega, \phi_\omega) = 1 \text{ for all } \omega.$$

By periodicity, for the Fourier basis functions we have

$$(e^{i\omega x}, e^{ivx}) = \frac{1}{i(v - \omega)}[e^{i(v-\omega)x}]_0^{2\pi} = 0, \qquad \omega \neq v,$$

i.e., they are orthogonal. To obtain orthonormality for the Fourier functions, we divide them by $\sqrt{2\pi}$, resulting in the formulas

$$f(x) = \frac{1}{\sqrt{2\pi}} \sum_{-\infty}^{\infty} \hat{f}_\omega e^{i\omega x},$$

$$\hat{f}_\omega = \frac{1}{\sqrt{2\pi}} \int_0^{2\pi} f(x)e^{-i\omega x} dx.$$

In the next section we shall discuss the Gibbs phenomenon, and in this connection we make a few comments on the convergence properties of Fourier series.

3.6.4 The Gibbs Phenomenon

It turns out that Fourier expansions are not good for approximating functions with low regularity, and for discontinuous functions they are really bad, which was discovered in the nineteenth century. In 1848, the English mathematician Henry Wilbraham (1825–1883) found that severe oscillations occurred around discontinuities, and even if they became less severe with an increasing number of terms in the expansions, there was always a substantial error close to the jump. However, his result seems to have passed without much notice, and it took about 50 years until the problem was brought to the attention of the science community.

Josia Willard Gibbs (1839–1903) got his doctorate in engineering in 1863 at Yale University with his thesis "On the form of the teeth of wheels in spur gearing". He had a very broad area of interests including physics, chemistry and mathematics, and in 1871 he became professor of mathematical physics at Yale. Later he became interested in the convergence of Fourier series, and in the two Nature articles [68] and [69] he brought up the difficulty in approximating discontinuous functions and the presence of erroneous oscillations. His first article actually had an error, and his second one was a correction to it. In 1906, the American mathematician Ferdinand Bôcher (1867–1918) gave a more precise analysis of the reason for the oscillations, which he called the *Gibbs phenomenon*.

An example of the phenomenon is shown in Fig. 3.4. It shows the result when approximating the function

$$f(x) = \begin{cases} \pi/4, & 0 < x \leq \pi, \\ -\pi/4, & \pi \leq 2\pi, \end{cases}$$

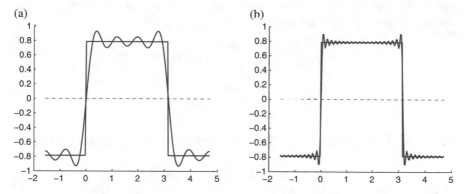

(a)

(b)

Fig. 3.4 Fourier series approximation of a discontinuous function (a) N = 4. (b) N = 20

which can be extended periodically to all x. The Fourier expansion is

$$g(x) = \sum_{j=1}^{N} \frac{1}{2j-1} \sin\big((2j-1)x\big),$$

and the figure shows f and g for $N = 4$ and $N = 20$, respectively.

This presence of the Gibbs phenomenon indicates that one should stay away from nonsmooth functions in general when using Fourier expansions. Actually, even without discontinuities, the accuracy of Fourier expansions is poor if f has low regularity, but one can show that the convergence rate with a growing number of terms becomes higher with increasing regularity. But even for discontinuous functions, the situation is not hopeless since there are ways to eliminate the Gibbs phenomenon, one of them being the addition of some smoothing operator which has a damping effect on the oscillations. The disadvantage of this technique is that even the discontinuities become smoothed out, and this may be a too severe effect. But there are other tricks that can be used, and in Sect. 5.7.2 one such trick is demonstrated when applied to computed tomography (CT).

3.6.5   The Fourier Transform

The limitation to periodic functions is of course restrictive. Sure enough, efforts to remove this restriction soon were tried. Clearly the series representation must be abandoned. By letting the length of the period tend to infinity, one approaches nonperiodic functions defined on the whole real axis. This means that discrete wave numbers ω transfer into a real variable ξ with $-\infty < \xi < \infty$. The function $f(x)$ and the Fourier transform are now represented as integrals

$$f(x) = \frac{1}{\sqrt{2\pi}} \int_{-\infty}^{\infty} \hat{f}(\xi) e^{i\xi x} d\xi,$$

$$\hat{f}(\xi) = \frac{1}{\sqrt{2\pi}} \int_{-\infty}^{\infty} f(x) e^{-i\xi x} dx.$$

Here it is assumed that $f(x)$ tends to zero sufficiently fast such that the integral exists for all ξ.

The continuous Fourier transform has played a fundamental role in the history of applied mathematics. But here we are concerned with computational aspects, and then we must discretize in some way. Fourier series can always be truncated after a finite number of terms, but the coefficients are still defined by integrals. The fully discrete algorithm can be described as follows.

Assume the function $f(x)$ is defined pointwise on a uniform grid $\{x_j = jh\}_j$ and that it is 2π-periodic. Using the notation $f_j = f(x_j)$, we assume that $Nh = 2\pi$

such that

$$f_{j+N} = f_j$$

for all j. The discrete Fourier transform (DFT) is defined by

$$\tilde{f}_\omega = \frac{1}{\sqrt{2\pi}} \sum_{j=0}^{N-1} f_j e^{-i\omega x_j} h, \qquad \omega = 0, 1, \ldots, N-1,$$

and the inverse transform gives the original grid function back:

$$f_j = \frac{1}{\sqrt{2\pi}} \sum_{\omega=0}^{N-1} \tilde{f}_\omega e^{i\omega x_j}, \qquad j = 0, 1, \ldots, N-1.$$

By defining $f(x)$ as

$$f(x) = \frac{1}{\sqrt{2\pi}} \sum_{\omega=0}^{N-1} \tilde{f}_\omega e^{i\omega x}, \qquad 0 \le x \le 2\pi,$$

one can interpret the DFT as the solution of an interpolation problem having point values $\{f_j\}_j$, called *trigonometric interpolation*. This means, among other things, that it can be used to approximate derivatives of the underlying function $f(x)$. Again, differentiation in Fourier space is obtained simply doing multiplication by $i\omega$.

Fourier himself didn't invent the DFT. Again we have a case where problems in astronomy were the source of new numerical methods. Since planet orbits are periodic, it was natural to use periodic functions for describing them. The French mathematician Alexis Clairaut was probably the first person to use the discrete cosine-transform, and Lagrange to use the sine-transform, both in the middle of the eighteenth century.

The discrete Fourier transform is very useful for spectral analysis from a theoretical point of view. The drawback is that the sum must be computed for N different wave numbers, and this takes $\mathcal{O}(N^2)$ arithmetic operations. Later in this book we shall come back to the invention of the Fast Fourier Transform, which is a fundamental speedup of the computation.

The different types of expansions described in this section are used for representing a certain function. One can see an expansion as a transform of the function, i.e., instead of representing the function in terms of its values, it is represented by its expansion coefficients. When it comes to wave propagation, it is obvious what the Fourier expansion means. Each coefficient is associated with a certain wave number, or a certain frequency if the original function is time-dependent. However, as is often the case in mathematics, a certain mathematical concept can be transferred and used for very different applications. The Fourier transform came to be a very powerful tool for analyzing differential equations, and was even more pronounced in the analysis of numerical methods for partial differential equations. This would have been hard for Fourier to imagine. We shall come back to the development of these methods for analysis later.

3.7   Ordinary Differential Equations (ODE)

With the introduction of differential calculus during the seventeenth century came the need to solve differential equations. However, analytic solutions can be found only for very simple problems, and it is necessary to construct methods that can produce approximate solutions. Discretization, such that the unknown variable is represented at grid-points, leads to finite difference methods. Since at the time hand calculations were the only possibility, one had to limit efforts to ordinary differential equations (ODE). Furthermore, the use of difference methods for boundary value problems for second and higher-order ODE lead to algebraic systems of equations, and hand calculation is not possible for solving any problems of practical interest. Therefore, the efforts were concentrated on initial value problems of the type

$$\frac{du}{dt} = f(t, u),$$

$$u(0) = u_0,$$

where in general f is a nonlinear function of its arguments. Higher- order ODE can always be rewritten as first-order systems. Problems in astronomy dominated, for example predicting the position of a planet at a certain time, and therefore we use t as notation for the independent variable.

As for many other problems, attempts were made quite early to use polynomials where the coefficients were determined in a recursive manner by using differential equation. However, it is not a very convenient method, and in the eighteenth century finite difference methods began to occur. These methods soon came to dominate this problem area, and they are still by far the most common methods for initial value problems for ordinary differential equations. In this chapter we shall describe how they were introduced and further developed.

3.7.1   The Euler Method

We have seen in the previous sections, as we will see later in this book, that new fundamental principles for numerical solutions have been invented by great mathematicians whose names will be well known far into the future. In the case of differential equations, the introduction of a numerical solution by difference methods is associated with Leonhard Euler. In [50], published in 1768, he introduced the idea of computing the solution stepwise, thus obtaining the solution at discrete points t_n. These points define a grid, and with a fixed stepsize h, we obtain a uniform grid

$$t_n = nh, \qquad n = 0, 1, \ldots.$$

Euler's method can be derived starting from the definition of the derivative:

$$\lim_{h \to 0} \frac{u(t+h) - u(t)}{h} = f(t, u(t)).$$

By choosing a finite but small h, and introducing the notation v_n for approximating $u(t_n)$, we get

$$v_{n+1} = v_n + hf(t_n, v_n), \qquad n = 0, 1, \ldots.$$

Assuming that the function f can be evaluated at each grid-point, this is a simple computation for each step. The principle of difference methods that later came to be so important for many applications had been introduced. We have again an example of a common technique for numerical computation: a very difficult problem involving complicated operators, like the differential operator in this case, is substituted using an algorithm containing simple algebraic operations.

The Euler method for ordinary differential equations can also be derived from a Taylor expansion

$$u(t+h) = u(t) + h\frac{du}{dt}(t) + \mathcal{O}(h^2) \tag{3.16}$$

by neglecting the h^2-term. From this representation, one can derive an error estimate

$$|v_n - u(t_n)| \le \text{const } h, \tag{3.17}$$

for any t_n in any fixed time interval $[0, T]$.

From a modern point of view, the method is very crude. A reasonable accuracy at a certain time $t = T$ requires a small stepsize h, i.e., many computational steps must be carried out. The estimate (3.17) shows that the method has first-order accuracy, which is too low for effective computations. But the important point here is that a new type of computational method had been introduced.

The Euler method is the simplest example of an *explicit method*. An *implicit method* is characterized by the presence of $f(t_{n+1}, v_{n+1})$ in the formula for computing v_{n+1}. The simplest implicit method is

$$v_{n+1} = v_n + hf(t_{n+1}, v_{n+1}).$$

This method is actually called the Euler backward method, even if Euler himself didn't use it.

There is actually a third way of interpreting the Euler method. We have

$$u(t_{n+1}) - u(t_n) = \int_{t_n}^{t_{n+1}} \frac{du}{dt}(t)\,dt = \int_{t_n}^{t_{n+1}} f(t, u(t))\,dt, \qquad (3.18)$$

and if f is approximated by its value at $t = t_n$, then we get the Euler method. Approximating solutions to differential equations has been replaced by approximating integrals.

Euler (1707–1783) is one of the greatest mathematicians ever. He was born in Basel, where he also got his basic education while living with his grandmother. In agreement with most other world famous scientists, he started his university studies very early in life, in this case at the University of Basel at age 13. Three years later he had completed a thesis in philosophy, but he soon switched to mathematics and physics. His father was a friend of the well-known mathematician Johann Bernoulli, whose son Daniel had a position at the physiology department at the Russian Academy of Sciences in St. Petersburg. When leaving this post he recommended Euler as his successor since, strangely enough, Euler's attempts to become a professor in physics at the University of Basel were refused. He accepted the offer and moved to St. Petersburg in 1727, and stayed there until 1741 when he moved to the Berlin Academy. In 1766 he returned to St. Petersburg where he stayed until his death.

His research had an enormous breadth, and he came up with new mathematical models for many different physical problems. One of his best known achievements involves the Euler equations, a set of partial differential equations describing the dynamics of a fluid. These equations are the basis for countless computer programs, some of them used when trying to overcome serious difficulties caused by nonlinearities which give rise to discontinuities of the true solutions.

A remarkable fact is that Euler was the father of much of the mathematical notation that we use today. Examples are writing $f(x)$ for a function of the independent variable x, the notation $\sin x$, etc. for trigonometric functions, the letter e for the base of the natural logarithm, the letter i for the imaginary unit, and the Greek letter Σ for summation. It is somewhat curious that one person had such an influence that all these notations survived and seem to be established forever. The newly published book [19] is an extensive Euler biography with a thorough description of his life and work.

Leonhard Euler
(Bernoulli-Euler-Zentrum
Basel)

3.7.2   Adams Methods

Not much happened for a long time when it comes to the further development of numerical methods for differential equations. Indeed, it was not until the second half of the nineteenth century that any significant efforts were made, and this took place in England.

The main problem with the Euler method is the low-order of accuracy. The error in the numerical solution is of the same order as the stepsize h, and for any reasonable accuracy, it requires a small h, which leads to a large number of grid-points where $f(t_n, v(t_n))$ must be evaluated. The obvious remedy is to construct a method where the order of the error is proportional to h^p, $p > 1$. But how should it be done without introducing so much extra work that the gain in accuracy does not pay off? One natural generalization of the Euler method could have been to extend the Taylor expansion (3.16) such that it would include more terms. In this way there would be a potential to obtain higher-order accuracy, since the remainder in the expansion is of a higher-order in h. The obvious drawback is that higher-order derivatives $d^j u/dt^j$ become involved, and they cannot be computed by direct evaluation of $f(t, u)$.

John Couch Adams (1819–1892) at the University of Cambridge was an astronomer, and is best known for predicting the existence and position of the planet Neptune. This remarkable achievement was not based on a direct observation of the planet through a telescope. Instead Adams observed the track of the known planet Uranus and used Newton's and Kepler's differential equations governing the motion. In this way he found that the computed track did not agree exactly with the observations, and his conclusion was that there must be another planet, which was confirmed by the first observation at the Berlin Observatory in 1846. Actually, the same type of prediction had been made by Urbain Jean-Joseph Le Verrier in Paris, and there is a dispute as to who really should have the credit.

John Couch Adams
(Photographer unknown,
public domain. Wikimedia*)

For the purpose of this book, Adams is the primary man of interest when it comes
to ordinary differential equations. The reason is that he worked out new numerical
methods for solving ODE, and these methods are still the basis for many standard
ODE-solvers. So here we have again a striking example in which new numerical
methods were developed as a result of the need to solve new problems in another
science, in this case astronomy as in so many other cases.

So, what was the new idea? Euler used only the previous value v_n when computing v_{n+1}, but Adams introduced more points back in time when he constructed his
methods. The first generalization is

$$v_{n+1} = v_n + \frac{h}{2}\big(3 f(t_n, v_n) - f(t_{n-1}, v_{n-1})\big), \tag{3.19}$$

which is a second-order method, i.e.,

$$|v_n - u(t_n)| \le K h^2,$$

where the constant K is independent of h. The method (3.19) is a *linear multistep
method*, which has the general form

$$v_{n+1} + \alpha_0 v_n + \ldots + \alpha_{k-1} v_{n-k+1}$$
$$= h\big(\beta_{-1} f(t_{n+1}, v_{n+1}) + \beta_0 f(t_n, v_n) + \ldots + \beta_{k-1} f(t_{n-k+1}, v_{n-k+1})\big), \tag{3.20}$$

where α_j, β_j are constants with $\alpha_{k-1}\beta_{k-1} \ne 0$. This is a k-step method, and if
$\beta_{-1} = 0$, it is explicit, otherwise implicit. The label *linear* does not refer to the
function f but rather to the fact that the formula (3.20) is a linear combination of v_j
and the functions $f(t_j, v_j)$.

Adams didn't work with the general form (3.20), but restricted his methods to the special case with coefficients on the left-hand side satisfying

$$\alpha_0 = -1,$$

$$\alpha_1 = \alpha_2 = \ldots = \alpha_{k-1} = 0.$$

The coefficients β_j are then chosen so that the accuracy becomes as high as possible, possibly with the extra condition that $\beta_{-1} = 0$ such that the method becomes explicit.

It took a long time before Adams methods were published. He had a colleague Francis Bashforth (1819–1892) at Cambridge who was interested in capillary action including the form of fluid drops, and in 1883 he published his new mathematical model for that problem in the book [5] with Adams as a coauthor. The title page says *An attempt to test the theories of capillary action by comparing the theoretical and measured forms of drops of fluid, with an explanation of the method of integration employed in constructing the tables which give the theoretical forms of such drops.* Adams's contribution is a whole chapter with a long title beginning *On the calculation of the theoretical forms of drops...* , and there he derives a general form of explicit multistep methods. This is then applied to the differential equation derived by Bashforth for drop formation. Both authors refer to each other in the text, but it is clear that Bashforth didn't have any role in the basic construction of the numerical method. However, this publication is the reason for the name Adams–Bashforth methods, which is by now firmly established for explicit Adams methods.

Implicit methods were also constructed by Adams, and he used Newton's method for solving the nonlinear equation obtained at $t = t_{n+1}$. Strangely enough, even these methods got an extra name when referring to them in later literature. Forest Ray Moulton (1872–1952) was an astronomer at the University of Chicago. When solving certain problems concerning celestial bodies he used a combination of the explicit Adams methods and an implicit method in the predictor-corrector sense. This means that an explicit method was used to make a preliminary prediction of the solution $\tilde{v}_{n+1}$ which was then used for evaluating $f(t_{n+1}, \tilde{v}_{n+1})$ in the implicit method. This particular procedure was published in 1926, see [122], and this was the reason for the name Adams–Moulton methods for the class of implicit methods that Adams had invented much earlier.

There is of course a drawback with multistep methods. There is only one initial value $u(0)$ given to define a unique solution $u(t)$ for all $t \geq 0$. Therefore, the $k - 1$ extra initial values that are required to get the algorithm going must be computed by some other numerical method. For a second order two-step method, the Euler method is good enough, since after just one step it has a local accuracy of order h^2. But for higher order methods, other initial methods must be used.

An observation can be made here. Adams used only the two points t_{n+1} and t_n on the left-hand side of (3.20), and this is of course a restriction. As an example, the formula can be centered at t_n, giving the second-order method

$$v_{n+1} = v_{n-1} + 2hf(t_n, v_n).$$

This method is called the *leap-frog scheme* because of its structure. It is a simpler and faster method, but not much used. We have not yet discussed any stability issues, but we shall see in Sect. 4.2 that stability problems restrict the leap-frog scheme to very special differential equations.

3.7.3 Runge–Kutta Methods

We have already noted that the development of numerical solution methods for ODE was quite slow after the initial attempt by Euler. At Cambridge University the topic was revived by Adams, but it is not clear what impact it had outside of Britain, in particular in Germany. Carl Runge, mentioned in Sect. 3.2, began working on it at the end of the century, and he began his article [142] published in 1895 by noting

> The calculation of any sort of numerical solution of a given differential equation whose ana-lytical solution is not known appears to have excited little interest among mathematicians up till now,.... (Translation from German).

Runge started out from the representation (3.18), and saw the problem as a question of how to approximate the integral. Such methods existed, and Simpson's rule was well known, but a difficulty remains. If the solution is known at $t = t_n$, the next step is to compute it at $t = t_{n+1}$, and if the integration formula involves $v(t_{n+1})$, a nonlinear equation must be solved, and we have an implicit method. Runge found a way to circumvent this difficulty by computing successive approximations starting from $v(t_n)$, thereby avoiding the involvement of v at earlier grid-points. This technique was further developed and systemized by Karl Heun (1859–1929) in his article [92] while working in Berlin.

The third German working on the numerical solution of ODE was Martin Wilhelm Kutta (1867–1944) at the University of Munich (later at the University of Stuttgart). Inspired by the articles of Runge and Heun, he developed what is today called the class of Runge–Kutta methods. (There are now three concepts including Runge in their name: Runge function, Runge phenomenon, Runge–Kutta methods.) The main principle is again to obtain higher-order accuracy by successively evaluating $f(t, v)$ for different values of t and v. The classical and most famous version is the 4th order method defined as follows:

$$k_1 = hf(t_n, v_n),$$

$$k_2 = hf(t_n + \frac{h}{2}, v_n + \frac{k_1}{2}),$$

$$k_3 = hf(t_n + \frac{h}{2}, v_n + \frac{k_2}{2}),$$

$$k_4 = hf(t_n + h, v_n + k_3),$$

$$v_{n+1} = v_n + \frac{1}{6}(k_1 + 2k_2 + 2k_3 + k_4).$$

The method is called a four stage method, because the computation for each time-step requires four intermediate stages with separate evaluations of f. However, it is still called a one-step method in analogy with the Euler method. The reason is that it can be written in the form

$$v_{n+1} = Q v_n,$$

where Q is an operator that assigns a unique value of the new approximation v_{n+1} in terms of the previous approximation v_n. As a consequence, we can get the algorithm going from the start, without any extra initial conditions in addition to the exact one $v_0 = u_0$.

Perhaps a little surprising, it took a long time until the very important question of stability was considered. It seems that Henri Poincaré (1854–1912) was the first to express concern about this topic. The stability of the solar system was of prime interest at this time, and he wrote in [135] published in 1892:

> Moreover, certain theoretical consequences that one might be tempted to draw from the form of these series are not legitimate, on account of their convergence. It is for this reason that they cannot be used to resolve the question of the stability of the solar system. (Translation from French.)

Even if he referred to series expansion, his remarks could as well be applied to finite difference methods. For the simple differential equation

$$\frac{du}{dt} = -\alpha u,$$

where α is a positive constant, stability and convergence for the Euler method requires that the amplification factor $1 - \alpha h$ is less than one in magnitude. After the introduction of electronic computers the question of stability gained much more interest since one could now carry out a very large number of time-steps. This will be discussed in Sect. 4.2.

3.7.4 Richardson Extrapolation

Linear multistep methods as well as Runge–Kutta methods require more work per step for higher-order accuracy compared to lower order, and this is of course no surprise. As in everyday life, it is hard to get anything for free. On the other hand, with a given error tolerance one can choose a larger stepsize with a higher order method, and the total work may be less. We should again remember that no computers were available, and in order to obtain any useful results for interesting problems, the work really had to be minimized.

Before going into ODE-solvers, we go back again to the ever interesting problem of computing π, and to the Dutch mathematician Cristian Huygens (1629–1695). He considered Archimedes method of inscribing a polygon with N equal sides of length h in the circle with radius 1 and then computing the total length p_N of this polygon. Then it is easy to show that the true arc-length $h = 2\pi/N$ of each circle sector differs from the cord by a quantity of order h^3, which gives a total error of order h^2. So, assuming that there is an expansion,

$$2\pi = p_N + ch^2 + \mathcal{O}(h^4), \tag{3.21}$$

where c is a constant which is independent of h. The same expansion holds also for the previous step with $N/2$ sides such that

$$2\pi = p_{N/2} + c(2h)^2 + \mathcal{O}(h^4).$$

By multiplying the first expansion by 4 and subtracting the second we obtain

$$2\pi = \frac{4p_N - p_{N/2}}{3} + \mathcal{O}(h^4),$$

and this is the formula that Huygens recommended. Without going any further in refining the polygon approximation and instead combining the two existing results, he obtained an approximation of π with an error of order h^4.

Much later the same problem was considered by M. Saigey with a generalized form of the same idea. In 1859 he published the book [147], and in order to illustrate the voluminous description of a method in words often used at this time, we cite the exact text.

> From a theorem which is our own, if we consider the polygons of the preceding number as the first approximations of the circumference, we shall obtain second approximations by adding to each of the first ones the third of its excess over the preceding one; third approximations, by adding to each of the second approximations the 15th of its excess over the preceding one; fourth approximations by adding to each of the thirds the 63th part of its excess over the preceding one; and so on indefinitely, the fractions 1/3, 1/15, 1/63, … having as denominators, respectively, $(4 - 1)$, $(4^2 - 1)$, $(4^3 - 1)$, …. (Translation from French.)

Obviously this long sentence describes the method obtained by including more terms in the series expansion (3.21), where the h-power terms are eliminated one-by-one.

Lewis Richardson (1881–1953) was a British mathematician who had strong interest in various applications, in particular meteorology; we shall come back to this in Sect. 3.8.4. He picked up the extrapolation idea described above, but he worked it out for more general problems involving the approximation of functions $f(x)$. In particular he showed how to improve the accuracy of difference methods for ODE. When introducing the true solution in a difference scheme, the truncation error can be expressed in a Taylor series around some grid-point. This series is expressed in terms of the stepsize h and the derivatives of the solution $u(t)$. This is a measure of how well the difference formula approximates the differential equation. It is by no means trivial to prove that the error in the *solution* $v(h, t)$ can be expanded into a Taylor series of the same type, but under certain conditions it is possible. If, furthermore, the difference formula is symmetric around some point, the expansion contains only even order powers of h, and we have

$$u(t) - v(h, t) = \phi(t)h^p + \mathcal{O}(h^{p+2}),\tag{3.22}$$

where the function $\phi(t)$ is independent of h and p is even. For example, the Adams–Moulton implicit method

$$v_{n+1} - v_n = \frac{h}{2}\big(f(t_{n+1}, v_{n+1}) + f(t_n, v_n)\big),$$

also called the trapezoidal method, has such an expansion with $p = 2$.

Richardson's idea was to run the method for a certain stepsize h_1 and then run it a second time with the stepsize h_2. With $h_1 = h$, $h_2 = 2h$ and $p = 2$, the expansion (3.22) then gives the system

$$u(t) = v(h, t) + \phi(t)h^2 + \mathcal{O}(h^4),$$
$$u(t) = v(2h, t) + 4\phi(t)h^2 + \mathcal{O}(h^4),$$

and by eliminating the first error term just as above, we get

$$u(t) = \frac{4v(h, t) - v(2h, t)}{3} + \mathcal{O}(h^4).\tag{3.23}$$

If for N even, the grid $t_0, t_1, t_2, \ldots, t_N$ corresponds to the stepsize h, the more accurate result is obtained on the grid $t_0, t_2, t_4, \ldots, t_N$.

When the first error term is eliminated, the next term of order 4 can be eliminated in the same way by picking a third step-size, and then one can continue further. The resulting coefficients are those obtained by Saigey for calculating π as described above.

The method goes under the name Richardson extrapolation without any reference to Huygens and Saigey. It can be applied to other types of problems as well where there is an expansion of the error, for example numerical evaluations of integrals.

Richardson published his method in 1911, see [138], where he actually used difference methods for *partial* differential equations, see Sect. 3.8.4. However, his method did not attract much interest until much later when computers became available.

Richardson's life story is quite different from most scientists of this era. As a boy he was sent to a Quaker boarding school, and this made him become a pacifist, so that he was exempted from military service during the first world war. However, he voluntarily joined the Friends' ambulance unit attached to a French infantry division, where he worked for 3 years during the war. But his status as a conscientious objector made it difficult for him to obtain any regular university position despite his high level of scientific competence. While natural science and technology usually get strong boosts during big wars, Richardson was careful not to contribute to military projects. Indeed he resigned from his position at the Meteorological Office when it became part of the Air Ministry which handled the Royal Air Force. He was also active in the peace movement by developing new mathematical models for the purpose of understanding how conflicts arise.

Lewis Richardson   (U.S.
national oceanic and
atmospheric administration,
public domain. Wikimedia*)

3.8   Partial Differential Equations (PDE)

Accurate modelling of many physical and mechanical processes requires partial differential equations. At the end of the nineteenth century there was an abundance of such models for various problems. They contributed a lot to the understanding of the properties, but explicit solutions of the differential equations were of course impossible to find, except for very simple applications. Approximate numerical methods had to be constructed, and difference methods were the immediate choice as a generalization of the difference methods for ordinary differential equations. But the challenge was enormous. A discretization in both space and time giving any reasonable accuracy required such a large computational effort, that it was virtually impossible to realize it for any practical application. However, some attempts were still made, and we will describe them in this section.

Another fundamental approximation principle was the use of a combination of known functions such as polynomials. It had earlier been used for interpolation, where the simple principle is to match polynomials exactly to the data at a finite number of discrete points. The same approach could be used for PDEs, but in this case the challenge was to construct some convenient approximation principle. We begin by describing the early development of such methods.

3.8.1 The Ritz–Galerkin Method

Walther Ritz (1878–1909) was a Swiss theoretical physicist working in Göttingen. Among many other problems he considered the problem of finding the deformation of an elastic plate caused by a given force. The model

$$\frac{\partial^4 u}{\partial x^4} + 2\frac{\partial^4 u}{\partial x^2 \partial y^2} + \frac{\partial^4 u}{\partial y^4} = F(x, y), \qquad (x, y) \in \Omega \tag{3.24}$$

with proper boundary conditions on u was well known at the time, but there were no analytical solutions, and no methods for a numerical solution were available. Instead Ritz considered the integral

$$I(v) = \int\int_\Omega \left(\left(\frac{\partial^2 v}{\partial x^2}\right)^2 + 2\left(\frac{\partial^2 v}{\partial x^2}\right)\left(\frac{\partial^2 v}{\partial y^2}\right) + \left(\frac{\partial^2 v}{\partial y^2}\right)^2 - 2Fv \right) dx\, dy,$$

and formulated the problem as a *variational problem*:
Find the function u(x,y) such that

$$I(u) = \min_v \ I(v). \tag{3.25}$$

The connection between the variational integral formulation and the differential equation had already been considered by Euler. If the solution of the variational problem is sufficiently smooth, then with proper boundary conditions it would satisfy the differential equation (3.24).

The two formulations are not exactly equivalent as definitions of the function u. The difference is that the integral formulation does not require the same regularity conditions on the solution. It is cnough that the integral exists, which for example allows for the second derivatives to be discontinuous. The differential equation on the other hand requires that the derivatives of order-up-to four be well defined. We say that (3.25) is the *weak form* of (3.24).

The formulation as a minimization problem (or a more general variational problem) as the basis for computing the solution had been used by the English physicist Lord Rayleigh (1842–1919) in his work on finding eigen-frequencies of a vibrating string, see [136]. (He was awarded the Nobel Prize in Physics in 1904 for his discovery of argon.) Ritz's idea was to use the variational form for more general problems and define an approximate solution of the form

$$u_N(x, y) = \sum_{j=1}^{N} c_j \phi_j(x, y),$$

where the basis (or coordinate) functions ϕ_j are chosen so that they can be explicitly differentiated and integrated. This is called the *Rayleigh–Ritz method*. Polynomials are a natural choice, but they must be chosen so that the boundary conditions are satisfied. For example if Ω is the unit square and $u = 0$ on the boundary, Ritz suggested polynomials as basis functions, all containing the factors $1-x^2$ and $1-y^2$.

When introducing u_N into the expression to be minimized, and differentiating with respect to the coefficients c_j, the result is a linear system

$$Ac = b,$$

where A is an $N \times N$ matrix.

The treatment of this, and other problems using the same technique, was published in the articles [139] in 1908 and [140] in 1909. These articles certainly must be considered as some of the most important in computational mathematics of all time. We shall see later how this solution principle is the basis for the most frequently used present day numerical methods.

Ritz had already died in 1909 at the early age of 31. In the following years, his method was not much recognized anywhere except in Russia, where Boris Galerkin (1871–1945) was quick to pick up the idea. Galerkin had a quite interesting life history. He was born in 1871 in Polotsk, which belonged to the Russian empire at the time, but is now part of Belarus. He got his higher education in Minsk, and was then enrolled at the Petersburg Technological Institute where he graduated in 1899. He then worked as an engineer at the Kharkov Locomotive Plant and later at the Northern Mechanical and Boiler Plant.

Boris Galerkin   (Source: Russian Academy of Science)

He was an early active politician in the social democratic party and organized a union for engineers. For this he was arrested and put in prison (1907) for one and a half years. But there he had quite good working conditions, and he wrote a long and important scientific paper while incarcerated there. After getting released, he lost interest in political activities, and devoted the rest of his life to science and engineering.

It was natural for him to go back to the Petersburg Technological Institute, where he started teaching in 1909. His earlier work as an engineer was largely directed towards constructing buildings and bridges, and this led him to develop new methods for solving the differential equations connected to structural mechanics.

In his article [60] Galerkin generalizes the Ritz method and applies it to many different engineering problems. Furthermore, he showed how the method can be defined without going via the minimization formulation. The procedure is based on the same principle as indicated above when connecting differential equations to an integral formulation. Actually, we don't need any minimization formulation to derive it. Indeed, by starting from the differential equation $Lu = F$, we multiply it by a test function v, integrate the resulting equation over the domain Ω and finally require that the integrated equation hold for all functions v in a certain function space $\mathscr{S}$. Using Green's theorem, the smoothness requirements on u is reduced.

The reason for labelling the new form "weak" is twofold. The requirement on continuous derivatives up-to-some order is relaxed, and furthermore, the new equation need only hold in the integrated form.

The change of the differential equation to the integral form and then reducing $\mathscr{S}$ to a finite-dimensional space $\mathscr{S}_h$ is often called the *Galerkin method*. However, Galerkin himself refers to Ritz in his paper and calls the method the *Ritz method*. The proper label should be the *Ritz–Galerkin method*.

The most general formulation goes like this. We start out with the differential equation written in the form

$$Lu = F, \tag{3.26}$$

where L is a differential operator acting in a function space which includes the boundary conditions. The notation for the scalar product and the norm is (u, v), $\|u\|^2 = (u, u)$ respectively. The standard L_2 scalar product in 2D is

$$(u, v) = \int \int_\Omega a(x, y)u(x, y)v(x, y)dx\, dy,$$

where $a(x, y) > 0$ everywhere in Ω. The weak form is obtained by multiplying (3.26) by a function v and integrating:

$$(Lu, v) = (F, v).$$

If L contains differential operators of at least the second order, the order is reduced using integration-by-parts as indicated above. However, the weak form is still well defined with L kept in its original form, and it is used as such for first-order differential equations. In either case, a function space $\mathscr{S}$ is defined, including the boundary conditions. The subspace $\mathscr{S}_N \subset \mathscr{S}$ is finite dimensional and spanned by N basis functions $\{\phi_j\}_{j=1}^N$. The numerical solution

$$\sum_{j=1}^N c_j(\phi_j) \tag{3.27}$$

is then defined as the solution to

$$(Lu_N - F, \phi) = 0, \qquad \text{for all } \phi \in \mathscr{S}_N,$$

where integration-by-parts is carried out if L is a differential operator of order 2 or higher. The interpretation of this relation is that the residual is orthogonal to the subspace $\mathscr{S}_N$. The solution can be found by solving the finite-dimensional problem

$$\sum_{j=1}^{N} (L\phi_j, \phi_k)c_j = (F, \phi_k), \qquad k = 1, 2, \ldots, N$$

for the coefficients c_j.

We note that the method can be seen as a generalization of the least squares method described in Sect. 3.4. For example, for the case of function approximation, L is simply the identity operator and the subsequent machinery is the same. For the case of an overdetermined system of equations, u is a vector, L is a rectangular matrix and $(\cdot, \cdot)$ denotes the usual vector scalar product.

An excellent presentation of the history around the Ritz–Galerkin method is found in [61].

3.8.2   Courant's Article on FEM

If one is asking scientists and engineers what is a typical representative for scientific computing, there is a good chance that the answer will be the finite element method (FEM). No separate class of methods among all numerical techniques have probably had such an impact on so many applications. This is partly due to the fact that partial differential equations are such a common and important mathematical model, and partly that FEM has such an excellent potential for solving very complicated problems in practical engineering and scientific applications. Even if the geometry of the computational domain is very irregular with corners and curved boundaries, finite elements can be made very flexible, such that the accuracy of the approximation is retained. This is in contrast to finite difference methods, which are most naturally defined on regular and uniform grids, and therefore require more effort and modifications to fit the boundaries.

For elliptic problems, the basis for FEM is the variational form of the problem and Ritz–Galerkin methods as described above. For the simplest form of the Poisson equation

$$-\frac{\partial^2 u}{\partial x^2} - \frac{\partial^2 u}{\partial y^2} = F, \qquad u \in \Omega,$$

$$u = 0, \qquad u \in \partial\Omega,$$

the associated quadratic functional is

$$I(v) = (Lv, v) - 2(F, v)$$

with $L = -(\partial^2/\partial x^2 + \partial^2/\partial y^2)$ and the scalar product

$$(u, v) = \int\int_{\Omega} u(x, y)v(x, y)dxdy.$$

With finite elements it becomes important to make sure that the approximating functions are regular enough so that the integrals exist. This restriction on the functions is relaxed by integrating-by-parts using Green's formula:

$$(Lv, v) = -\int\int_{\Omega} \left(\frac{\partial^2 v}{\partial x^2} + \frac{\partial^2 v}{\partial y^2}\right)v\, dxdy = \int\int_{\Omega} \left(\left(\frac{\partial v}{\partial x}\right)^2 + \left(\frac{\partial v}{\partial y}\right)^2\right)dx\, dy.$$

The admissible functions v are now such that $v = 0$ on the boundary $\partial\Omega$, and

$$\int\int_{\Omega} \left(\left(\frac{\partial v}{\partial x}\right)^2 + \left(\frac{\partial v}{\partial y}\right)^2 + v^2\right)dxdy < \infty.$$

This function space is called H_0^1, and continuity of v is enough to make sure that the integrals exist.

Richard Courant (1888–1972) was a German mathematician with a Jewish background. Even if he had served in the German army during the first world war, he felt pressure from the Nazis and left Germany in 1933. After a year at Cambridge he accepted a position at New York University, where he soon founded an institute for applied mathematics. In 1964 it was renamed The Courant Institute of Mathematical Sciences, and is one of the most prestigious research institutes in the world.

Richard Courant   (Photo:
Konrad Jacobs; Source:
Archives of the
Mathematisches
Forschungsinstitut
Oberwolfach.)

In 1943 Courant published a paper on methods for certain variational problems, see [27]. He discussed the Rayleigh–Ritz method, in particular, the difficulties in choosing the basis functions ϕ_j for the approximating subspace. He has a special paragraph titled *Objections to the Rayleigh-Ritz method*. There he says: *"The vagueness as to the accuracy of the approximation obtained is only one of the objections to the Rayleigh-Ritz method that may be raised. More annoying is that a suitable selection of the coordinate functions is often very difficult and that laborious computations are sometimes necessary. For these reasons, alternative methods must be studied."*

After some comments on the superiority of "the method of differences" over the Rayleigh–Ritz method, he takes a completely new approach by suggesting that "the method of finite difference can be subordinated to the Rayleigh–Ritz method". This means that a rectangular grid is constructed, each rectangle is divided into two triangles by drawing a diagonal, and finally a linear function is defined on each triangle. The integrals are now easily computed in terms of the unknown coefficients, and the system of equations giving the minimum is solved.

This piece of the article does not contain a single equation or formula. However, it is still the introduction of the finite element method. In an appendix, the method is applied to a plane torsion problem for a square with a square hole in it with area A_0. The functional to be minimized is

$$D(\phi) = \int \int \left(\left(\frac{\partial \phi}{\partial x} \right)^2 + \left(\frac{\partial \phi}{\partial y} \right)^2 + 2\phi \right) dx\, dy.$$

On the outer boundary we have the condition $u = 0$. On the inner boundary Γ_0, there is a "natural boundary condition"

$$\int_{\Gamma_0} \frac{\partial u}{\partial n} ds + u_0 A_0 = 0,$$

where u_0 is an unknown constant value of u along the inner boundary.

The system of equations for the unknown coefficients of the polynomials is never written down, but the result is presented for four discretizations, the finest with 9 unknowns. Figure 3.5 shows this triangularization.

By symmetry, the computation can be limited to one eighth of the domain. There are only 8 nodes, but the 9th unknown is u_0 at the inner boundary.

This paper should be one of the most important in the history of numerical methods. However, it got little attention at the time and it was not until 10 years later that FEM got its real momentum. And then few people referred to the Courant paper as the source of the method.

The finite element method will be further discussed in Sect. 5.6.

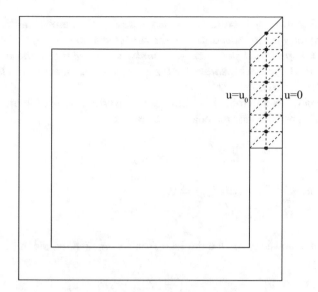

Fig. 3.5 The Courant torsion problem with 9 unknowns

3.8.3 Richardson's First Paper on Difference Methods

The first serious attempt to solve a PDE-problem numerically was made by Lewis Richardson (already mentioned in Sect. 3.7.4). He was among the first to use finite difference methods for PDE, and his article [138] on this method is a remarkable one. He not only describes the principles both for initial value and boundary value problems in a quite general manner, but he also has a number of practical and philosophical remarks. His formulation in the introduction of the paper is worth noting: *"Finite differences have, in themselves, but little importance to the student of matter and ether. They are here regarded simply as a makeshift for infinitesimals; and the understanding is always that we will eventually make the differences so small that the errors due to their finite size will be less than the errors of experiment of practical working, and may therefore be disregarded."* (All citations of Richardson in this section can be found in [44].)

Later, when he discusses boundary value problems leading to large systems of algebraic equations, he notes that the labor becomes very great *"and is of a sort which a clerk will not easily do"*. Accordingly, he must hire skilled people that he calls computers, and he organizes the computations in a very structured way. Furthermore he pays them well according to their achievements. For example, for the computation of the difference expression

$$v_{j+1,k} + v_{j,k+1} + v_{j-1,k} + v_{j,k-1} - 4v_{j,k},$$

he pays $n/18$ pence per coordinate point, where n is the number of digits. He also comments on the working speed: *"As to the rate of working, one of the quickest boys averaged 2000 operations per week, for numbers of three digits, those done wrong being discounted, the weekly salary for the quick boy was one pound and 8 shilling. No pay for erroneous numbers"*.

Richardson treats a number of examples in his article. One of them is the initial boundary value problem for the heat equation

$$\frac{\partial u}{\partial t} = \frac{\partial^2 u}{\partial x^2}, \qquad -0.5 \le x \le 0.5, \;\; 0 \le t,$$
$$u(-0.5, t) = u(0.5, t) = 0, \tag{3.28}$$
$$u(x, 0) = 1.$$

He uses what is nowadays called the leap-frog scheme mentioned in Sect. 3.7.2

$$v_j^{n+1} - v_j^{n-1} = \frac{2\Delta t}{\Delta x^2}(v_{j-1}^n - 2v_j^n + v_{j+1}^n), \tag{3.29}$$

which is centered at the grid-point $(x_j, t_n) = (j\Delta x, n\Delta t)$. Here we have used the notation v_j^n for $v(x_j, t_n)$. Richardson says in his paper *"The method of this example is so simple that it can hardly be novel"*. It is true that it is simple, but it is still probably new since anybody using it for any realistic example would find that it is not good for a parabolic PDE such as the heat equation. The method is unstable; we shall come back to this topic in Sect. 4.3.1.

However, with a very small Δt compared to Δx a few initial timesteps may be taken without triggering the unstable mode of the solution. Sure enough, Richardson chose $\Delta x = 0.1$, $\Delta t = 0.001$, which seems odd since the method has second-order accuracy both in space and time. One can only speculate, but a qualified guess is that he started out with a much larger Δt, but had to decrease it substantially in order to get an accurate result after 5 steps.

But there is also another problem with the method. The leap-frog scheme is a two-step method, and it requires a one-step method to calculate the first level at $t = \Delta t$. Richardson uses what is nowadays called the Crank–Nicolson method (see Sect. 4.3.1)

$$v_j^{n+1} - v_j^n = \frac{\Delta t}{2\Delta x^2}(v_{j-1}^n - 2v_j^n + v_{j+1}^n + v_{j-1}^{n+1} - 2v_j^{n+1} + v_{j+1}^{n+1}), \tag{3.30}$$

which is centered at the non-existing point $(x_j, t_{n+1/2})$. The computation of this first step requires the solution of a system of algebraic equations coupling all unknowns v_j^1 to each other. Richardson is relieved after getting over this initial difficulty saying *"Having gotten over this rather troublesome first step, we can find the rest much more simply... "*. Ironically, he could have used $\Delta t = 0.005$ in the first step with the Crank–Nicolson method reaching the end-point in one step and achieved essentially the same accuracy (maximum error=0.026) as he got with the leap-frog scheme after 5 steps (maximum error=0.022).

In the same article, Richardson presents the solution of a boundary value problem describing stresses in a dam wall. The presentation is very detailed, and uses iteration for solving the linear system of algebraic equations that arise from the discretization of the PDE.

The article by Richardson is an impressive piece of work. He opened up new possibilities by showing how difference methods could be used for solving partial differential equations. With only human computers available he managed to obtain numerical results for problems that were nontrivial. Even if his article [138] didn't get very much attention at the time, his work was the basis for PDE-solvers during the enormous computational boost that occurred with the postwar introduction of electronic computers.

3.8.4   Weather Prediction; A First Attempt

Richardson had a broad spectrum of interests and knowledge. Among other things, he had an interest in meteorology, and he wanted to find methods for weather prediction. Many researchers (and others) had tried many different approaches, but no one had succeeded. One idea that had been tried by many was to compare the current weather pattern with similar ones from the past, and then assume that the weather during the coming days also would be similar to the weather of the subsequent days in the past. Richardson picked up the fundamentally different approach of using the differential equations governing the dynamics of the atmosphere. These equations were well known at the time; they are largely based on the Euler equations of fluid dynamics. An explicit difference method was used to compute the solution by stepping forward in time starting from a known state, a certain day years back, to obtain the solution 6 h later. Unfortunately, the result was completely wrong, despite the fact that the arithmetic contained no errors except for rounding errors. So, what went wrong?

The weather as we know it moves with moderate speed. A typical speed of a weather front with rain is 30 km/h. If the discretization in space is such that the computational points are 180 km apart, it seems reasonable to take a 6 h timestep in order to get an effect at a certain point from the neighboring point. There are of course variations of air pressure and other variables between the computational points, i.e., the grid is too course, but the main effects of the front should come out with a fair accuracy. However, there are several differential equations that are coupled to each other, and this allows for propagation of the gravity waves as well. These waves contain a certain energy and move considerably faster than the weather front. However, under reasonable conditions they have little influence on the fundamental variables that govern the weather, and therefore they can be disregarded.

One way to avoid the difficulty is to specify the initial conditions such that the gravity waves are never triggered. We use a trivial example to illustrate the mechanism. Consider the ordinary differential equation

$$\frac{d^2u}{dt^2} - \lambda^2 u = 0, \qquad 0 \le t,$$

to be solved as an initial value problem with $u(0) = 1$. The solution has the general form

$$u(t) = ae^{-\lambda t} + be^{\lambda t},$$

where $\lambda > 0$. We are interested in bounded solutions for all t, and in order to guarantee this, the second initial condition is prescribed as

$$\frac{du}{dt}(0) = -\lambda u(0) \tag{3.31}$$

to obtain the solution

$$u(t) = ae^{-\lambda t}.$$

The trick to get rid of the parasitic exponentially growing solution is to specify the condition (3.31). This is called *initialization*, i.e., we make sure that the initial data are such that certain potential components of the solution are eliminated.

Let us now assume that the initialization is not done properly, and we have

$$\frac{du}{dt}(0) = -\lambda(1 + \varepsilon)u(0),$$

where $|\varepsilon|$ is small. The solution is now

$$u(t) = ae^{-\lambda t} - \frac{\varepsilon}{2 + \varepsilon}e^{\lambda t},$$

i.e., the parasitic solution is activated and will after some time destroy the true solution completely.

The situation is of course much more complicated when it comes to meteorology, but the principle is the same. The gravity waves have little influence on the weather, and in order to avoid them, the initial state must be prepared in a certain way. However, the differential equations are nonlinear which means that there is no guarantee that the gravity waves stay silent. Furthermore, the difference method has truncation errors which introduce non-physical perturbations in the approximate solution. One type of cure is to introduce a filter that will damp out the fast varying components without affecting the main part of the solution. Indeed, that technique is used today, but it was not available to Richardson.

Finally, there is a fundamental restriction on the timestep for any explicit difference scheme. The fastest waves govern the choice of timestep, and for the meteorology problem this means that the timestep must be chosen much smaller than would be reasonable, rather than considering the typical speed of weather changes. But the theory for this choice was not available either. It came with an article in 1928 that became one of the most cited in the history of PDE-computations.

With today's knowledge, it is no surprise that Richardson's computation gave a result that had nothing to do with the correct weather that had been observed. But he was on the right track in realizing that a discretization of the proper differential equations was the only way to predict weather with any accuracy.

3.8.5   The CFL-Article

As mentioned in Sect. 3.8.2, Richard Courant was a German mathematician who moved to the United States in 1933. While still in Germany, he worked together with Kurt Otto Friedrichs (1901–1982) and Hans Lewy (1904–1988) at the University of Göttingen. In 1928 they published the remarkable article [28], later republished in English, see [30]. It is the first article that gives a general treatment of difference methods for PDEs, and goes into considerable theoretical depth.

For elliptic equations they consider the boundary value problem, and note that the difference approximation leads to a nonsingular linear system of algebraic equations that has a unique solution. This fact can be used to prove existence of a unique solution to the underlying PDE, which gives a new mathematical tool for analyzing differential equations. The same principle can be applied to the differential operator eigenvalue problem

$$Lu = \lambda u,$$

$$u = 0 \qquad \text{on the boundary,}$$

which is replaced by an algebraic eigenvalue problem with properties that are better known.

The article is the first one that contains a strict convergence proof for solving the difference approximation for solutions of the PDE as the stepsize tends to zero. This lay the groundwork for practical computations that materialized much later when electronic computers appeared. If a computation with a certain stepsize h seemed to give a result of too poor quality, one could be confident that a computation with the stepsize $h/2$ would give a better result

The article also contains a quite different numerical method for elliptic problems, namely the method of the random walk. This is quite an original approach, and is typical for these excellent mathematicians.

But the largest impact and most cited result of this article is the discussion of a hyperbolic PDE. In particular, the authors studied the initial value problem for the wave equation

$$\frac{\partial^2 u}{\partial t^2} = \frac{\partial^2 u}{\partial x^2},$$

$$u(x,0) = f(x), \tag{3.32}$$

$$\frac{\partial u}{\partial t}(x,0) = g(x),$$

and the difference scheme

$$u_j^{n+1} - 2u_j^n + u_j^{n-1} = \left(\frac{\Delta t}{\Delta x}\right)^2 (u_{j+1}^n - 2u_j^n + u_{j-1}^n),$$

where Δx and Δt are the space- and timesteps respectively. Figure 3.6 shows how u_j^{n+1} on the top is computed in terms of four values behind. The solution of the problem (3.32) is obtained from d'Alembert's formula

$$u(x,t) = \frac{1}{2}\left(f(x-t) + f(x+t) + \int_{x-t}^{x+t} g(\xi)d\xi\right).$$

The solution u at the point (x, t) depends on the initial data in the interval $\Omega = [x - t, x + t]$, which is called the *domain of dependence*. Consider now the computational grid and the difference scheme. If the stepsizes are chosen as constants and such that $\Delta t \leq \Delta x$, then at least all the grid-points inside the domain of dependence are located inside the *numerical domain of dependence* Ω_h including all grid-points on the initial line that contribute to the computing u_j^{n+1}. This is illustrated in Fig. 3.7, where $\Delta t = 2\Delta x/3$. If the grid is further refined, but with

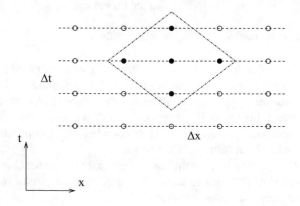

Fig. 3.6 Computational stencil

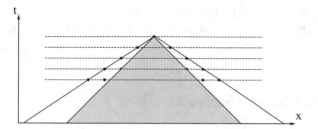

Fig. 3.7 Domain of dependence for differential equation and difference scheme

the same ratio between Δt and Δx, the grid-points in Ω are still included in Ω_h, and we can expect better accuracy as a result of better resolution. Theoretically, we get convergence as $\Delta x \to 0$.

If, on the other hand, the timestep is chosen such that $\Delta t > \Delta x$, then Ω_h becomes too small. Since some of the grid-points in Ω are not included in Ω_h, the correct solution can never be obtained no matter how far the grid is refined. This is the famous *CFL-condition* and $\Delta t / \Delta x$ is called the *Courant number*, or maybe it is more common as the *CFL-number*. With proper modifications, the condition can be generalized to all hyperbolic problems. It can be expressed in simple terms in the following way:

The numerical domain of dependence for an explicit difference approximation of a hyperbolic problem must include the domain of dependence for the differential equation.

The CFL-paper [28] actually deals with the fundamental concept of stability. The difference scheme above is stable if and only if $\Delta t \leq \Delta x$. If this condition is not satisfied, one cannot expect any reasonable results. This is what happened to Richardson as described in the previous section. His timestep violated the CFL-condition. With the introduction of electronic computers, the stability theory came to be a central theme for research for a long time, and we shall come back to this.

The CFL-article is a typical example of mathematical research. When it was published, nobody was in the neighborhood of making any practical computations for solutions of PDE. Except for trivial problems, it was simply too much work for any human beings to carry out all the algebra needed to reach the result. However, the three Germans still worked out the theory and studied the limit process as the step-size tends to zero. With the electronic computers a few decades later, their results played a fundamental role in all kinds of applications.

3.9 Optimization

Optimization problems occur in just about every field, not only in science and engineering but, perhaps even more frequently, in economics and neighboring areas like routing and scheduling. There is a great difference in the degree of difficulty between linear and nonlinear problems. In the first case, there is usually a unique solution, while in the second case one has to be content with locally optimal solutions.

3.9.1   The Fourier–Motzkin Algorithm

Many important optimization problems are linear, which means that the function to be optimized is linear as well as the side conditions. A natural quite general form of the problem is

$$\max f(x_1, x_2, \ldots, x_n) = \max(c_1 x_1 + c_2 x_2 + \ldots + c_n x_n),$$

$$a_{11}x_1 + a_{12}x_2 + \ldots + a_{1n}x_n \leq b_1,$$

$$a_{21}x_1 + a_{22}x_2 + \ldots + a_{2n}x_n \leq b_2,$$

$$\vdots \tag{3.33}$$

$$a_{m1}x_1 + a_{m2}x_2 + \ldots + a_{mn}x_n \leq b_m,$$

$$x_1, x_2, \ldots, x_n \geq 0.$$

It seems that Fourier was the first to derive a systematic method for finding the maximum in 1827, see [54]. However, his work went quite unnoticed, and it took more than 100 years until it was rediscovered by Theodore Motzkin, who presented it in his Ph.D. thesis at the University of Basel in 1936 with the title *"Beiträge zur Theorie der linearen Ungleichungen" (Contributions to the Theory of Linear Inequalities)*. The method goes as follows.

As a first step, we introduce a new variable x_{n+1} and write

$$f(x_1, x_2, \ldots, x_n) = x_{n+1},$$

or equivalently,

$$c_1 x_1 + c_2 x_2 + \ldots + c_n x_n - x_{n+1} \leq 0,$$

$$-c_1 x_1 - c_2 x_2 + \ldots - c_n x_n + x_{n+1} \leq 0.$$

In this way x_{n+1} denotes the possible values of f, and if we can find the maximum of x_{n+1}, the problem is solved. The idea is to add two new inequalities to the original set of inequalities, and then solve this set by eliminating variable-by-variable starting with x_1. As a preparatory step, we normalize every inequality where x_1 occurs, dividing by a_{j1} so that the coefficient for x_1 becomes 1 or -1. In this way we obtain one set S_1 of the form

$$x_1 + \sum_{k=2}^{n+1} a_{jk}x_k \leq b_j, \qquad j = 1, 2, \ldots, n_1,$$

a second set S_2

$$-x_1 + \sum_{k=2}^{n+1} a_{jk}x_k \leq b_j, \qquad j = n_1 + 1, n_1 + 2, \ldots, n_1 + n_2,$$

and a third set S_3

$$\sum_{k=2}^{n} a_{jk}x_k \leq b_j, \quad j = n_1 + n_2 + 1, n_1 + n_2 + 2, \ldots, n + 1,$$

which consists of the original inequalities which do not contain x_1. Here we have renumbered the equations, and kept the original notation a_{jk}, b_j. Each inequality in S_1 is added to each inequality in S_2 giving $n_1 n_2$ new inequalities without x_1, and when adding S_3 to these, we have a new and larger set of inequalities, but with one variable less. The procedure is now repeated for this new set, and finally we have only x_{n+1} left. The upper bound $(x_{n+1})_{\max}$ is the maximum value of f, and for free we also get the minimum value of f as $(x_{n+1})_{\min}$. In order to find the point where the maximum is attained, we go back to the previous systems with $x_{n+1} = (x_{n+1})_{\min}$ inserted.

Let us demonstrate the procedure for a trivial case:

$$\max(x_1 + 3x_2) \text{ on the triangle } \{x_1 + x_2 \leq 1, \ \ x_1 \geq 0, \ \ x_2 \geq 0\}.$$

We forget about the easy solution obtained by just looking at a drawing of the triangle, and apply instead the Fourier–Motzkin algorithm. After adding the two new inequalities and rearranging, we obtain

$$x_1 + x_2 \leq 1,$$
$$x_1 + 3x_2 - x_3 \leq 0,$$
$$-x_1 \leq 0,$$
$$-x_2 \leq 0,$$
$$-x_1 - 3x_2 + x_3 \leq 0.$$

The elimination procedure gives

$$x_2 \leq 1,$$
$$-2x_2 + x_3 \leq 1,$$
$$3x_2 - x_3 \leq 0,$$
$$0 \leq 0,$$
$$-x_2 \leq 0,$$

and after reordering and normalization

$$x_2 \leq 1,$$
$$x_2 - \frac{1}{3}x_3 \leq 0,$$
$$-x_2 + \frac{1}{2}x_3 \leq \frac{1}{2},$$
$$-x_2 \leq 0.$$

A new elimination sweep gives

$$\frac{1}{2}x_3 \leq \frac{3}{2},$$
$$0 \leq 1,$$
$$\frac{1}{6}x_3 \leq \frac{1}{2},$$
$$-\frac{1}{3}x_3 \leq 0,$$

and the final result is

$$0 \leq x_3 \leq 3.$$

By going back, we find that $x_1 = 0$, $x_2 = 1$ for $x_3 = (x_3)_{max} = 3$, which gives the solution

$$\max f(x_1, x_2) = f(0, 1) = 3.$$

If the original problem is not properly posed, it will show up during the process. If for example the inequalities are such that the domain is empty, an inequality of the type $1 \leq 0$ will come up at some stage. As a consequence, a successful complete procedure is a constructive proof that the problem is properly posed, which was of course realized by Fourier.

From a computational point of view, the algorithm is in general quite bad. In the worst case, the number of inequalities grows terribly fast since it is essentially squared at each step of the elimination. Still, there are certain applications where the structure has a special form such that the elimination process is reasonable.

3.9.2   Linear Programming and the War Effort

Leonid Kantorovich (1912–1986) worked at Leningrad University in the thirties when he started consulting for the Laboratory of the Plywood Trust. They had the problem of finding the best distribution of materials to different factories for optimal production. It could be formulated as a linear optimization problem with a number of side conditions. He found that many different problems in economics could be formulated in the same way. The method presented in the previous section was out of the question; the necessary work was way too heavy to carry out manually. At this time it was known that the maximum must be obtained at a corner of the given domain, but there were simply too many corners.

Kantorovich started working on this problem, and he laid the foundation for what is now called *linear programming*. Starting from a corner of the polyhedron, the idea is to find a path for marching further so that each new corner gives a value of the object function f that is not smaller than the previous one. He published his method in 1939 in the booklet [96], which was published by Leningrad University Press. The second world war began at the same time, and the work went quite unnoticed outside of the Soviet Union. Kantorovich himself became engaged by the Soviet Army and made significant contributions to the solution of many crucial challenges. One example is the horrific situation of the inhabitants of Leningrad during the siege of 1941–1944. The only transport to the city was across the Ladoga lake, and in winter time the frozen lake offered possibilities for transporting food and ammunition into the city. It was called the Road of Life, and the problem was to plan the route and distance between the vehicles for maximal transport volume without breaking the ice. Kantorovich, who at this time worked for the Military Engineering-Technical University, made fundamental theoretical calculations which became crucial for the success of the undertaking.

When the second world war broke out, the young Ph.D. student George Bernard Dantzig (1914–2005) at the University of California, Berkeley, took a leave of absence and started working for the U.S. Air Force Office of Statistical Control. The war effort on the American side contained of course many optimization problems, and Dantzig worked on a general method for linear problems. His new method was later called the *Simplex Method*, and it was based on the same principles that Kantorovich had used. The work was kept secret as part of the war effort, and it was not published until 1947. His book [34] is an excellent thorough description of the method. We shall here indicate how it works.

The problem (3.33) is first reformulated by introducing slack variables, such that the inequalities are made into equalities. An equation of the type

$$a_{j1}x_1 + a_{j2}x_2 + \ldots + a_{jn}x_n \leq b_j$$

is changed to

$$a_{j1}x_1 + a_{j2}x_2 + \ldots + a_{jn}x_n + x_{n+1} = b_j, \quad x_{n+1} \geq 0.$$

We assume that this has been done, and introduce the notation $\mathbf{c}$ and $\mathbf{x}$ for the vectors with n components c_j and x_j respectively. The problem now takes the form

$$\max_{\mathbf{x}} f(\mathbf{x}) = \max_{\mathbf{x}} \mathbf{c}^T \mathbf{x},$$

$$A\mathbf{x} = \mathbf{b},$$

$$\mathbf{x} \geq \mathbf{0},$$

where the $m \times n$ matrix A and the vector $\mathbf{b}$ defining the side conditions are given. A *feasible vector* $\mathbf{v}$ is a vector that satisfies the side conditions, and an *optimal feasible vector* is a vector that solves the optimization problem. A *basic feasible vector* is a feasible vector with at most m nonzero components. We now use the important theorem that says that there is at least one basic feasible vector that is an optimal feasible vector. This is the basis for the simplex algorithm.

The first step is to find a feasible vector where m components can be expressed in terms of the remaining $(n - m)$ components. We look for an $m \times m$ submatrix composed of m linearly independent columns of A such that we can express the corresponding subvector and f in terms of the remaining components. Assuming that the first m columns of A are linearly independent, we get for the example $n = 5$, $m = 3$

$$f = \tilde{c}_4 x_4 + \tilde{c}_5 x_5 + \tilde{d},$$

$$x_1 = \tilde{b}_1 + \tilde{a}_{14} x_4 + \tilde{a}_{15} x_5,$$

$$x_2 = \tilde{b}_2 + \tilde{a}_{24} x_4 + \tilde{a}_{25} x_5,$$

$$x_3 = \tilde{b}_3 + \tilde{a}_{34} x_4 + \tilde{a}_{35} x_5.$$

If we are lucky, $\mathbf{x}$ is a feasible vector also for $x_4 = x_5 = 0$, and then we can continue. Otherwise we have to try another vector.

If $\tilde{c}_4$ and $\tilde{c}_5$ both are negative, we cannot increase f anymore, and we have found the optimal solution. Otherwise one of the variables x_r on the right-hand side that has a positive coefficient $\tilde{c}_r$ in f is interchanged with one of the variables x_l on the left-hand side. The choice of x_r and x_l is made such that f gets the largest increase while at the same time all the variables stay nonnegative. This step is then repeated until all the coefficients $\tilde{c}_j$ of the right-hand side variables are negative.

In practice, the maximum is often found after a number of steps that is proportional to m, even if it is possible to construct examples in which the convergence rate is much worse. Furthermore, there are certain degenerate cases where special procedures must be applied to get the algorithm going. However, the simplex method is considered as one of the great computational inventions of all time. There is no Nobel Prize in mathematics, but Kantorovich, who was the first to use the basic principles for the simplex method, received the "Prize in economic sciences in memory of Alfred Nobel" in 1975.

It is quite interesting that both Kantorovich and Dantzig were heavily involved in the war efforts. It is another example of how the war demanded the development of new computational methods that later came to be of fundamental importance for a wide range of applications. (This should not be used as an excuse for starting a war!)

3.9.3 Nonlinear Optimization

When mathematical models go from linear to nonlinear, the degree of difficulty goes up dramatically no matter what kind of model we are dealing with. The optimization problem is no exception. We know of course that the maximum or minimum (if it exists) of a nonlinear function $f(x_1, x_2, \ldots, x_n)$ defined on the whole space is to be found among the points where the gradient is zero:

$$
\nabla f = \begin{bmatrix} \partial f/\partial x_1 \\ \partial f/\partial x_2 \\ \vdots \\ \partial f/\partial x_n \end{bmatrix} = 0. \tag{3.34}
$$

The definition of stationary points came as an immediate consequence of the calculus created by Newton and Leibniz. However, (3.34) is a nonlinear system of equations, and in general it is impossible to solve it analytically. For numerical computation there is the iterative Newton–Raphson method described in Sect. 3.1.2, which requires the solution of a linear system in each step. Hopefully we can capture all the local maxima and minima, and then extract the one that is the largest or the smallest.

Other more effective numerical optimization methods were developed later. One of them is the method of steepest descent which is based on a simple principle. The gradient ∇f defines a direction in the n-dimensional space. For a given point $\mathbf{x}^*$, the negative gradient $-\nabla f(\mathbf{x}^*)$ points in the direction of steepest downhill slope, and we can define a one-dimensional function $g(\xi)$ with $\xi = 0$ corresponding to $\mathbf{x} = \mathbf{x}^*$. The minimum value of this function is found for some $\xi^* > 0$, and the new gradient giving the search direction at the new point is computed. In this way an iterative method is obtained in the form

$$\mathbf{x}^{(j+1)} = \mathbf{x}^{(j)} - \lambda_j \nabla f(\mathbf{x}^{(j)}).$$

The distance λ_j is varying with each iteration in this version. It can also be considered as a stepsize, and as such it can be fixed which gives a simpler algorithm.

The steepest descent method was first published in 1909 by the Dutchman Peter Debye, see [39], but in the article he referred back to unpublished work by Riemann. Debye received the Nobel Prize in Chemistry in 1936.

As we have seen above for linear problems, the introduction of constraints on the variables x_j causes severe complications, and of course it does so also for nonlinear problems. But such problems were already considered by Lagrange for problems having the form

$$\min_{\mathbf{x}} f(\mathbf{x}),$$

$$g_1(\mathbf{x}) = g_2(\mathbf{x}) = \ldots, = g_m(\mathbf{x}) = 0,$$

where $\mathbf{x}$ is a vector with n components. In 1804 he published the book "*Leçons sur le calcul des fonctions*", where he introduced what is now called the Lagrange multiplier method. The Lagrange function

$$f(\mathbf{x}) + \sum_{j=1}^{m} \lambda_j g_j(\mathbf{x})$$

is formed, and then the stationary points are found by setting to zero all partial derivatives with respect to the n variables x_j and the m variables λ_j. The resulting nonlinear system with $n + m$ unknowns can be solved by the Newton–Raphson method. However, a stationary point is not necessarily an extremum of the function, and better methods were constructed in the postwar era.

Chapter 4
The Beginning of the Computer Era

Mathematical modelling was well established as a scientific branch at the time of the second world war. However, it is one thing to develop a mathematical model containing equations describing a physical process; it is quite another to find a solution that satisfies the equations. The most common type of mathematical models among all physical problems is the differential equation, and most often it has several independent variables, i.e., it is a partial differential equation. Over the centuries, mathematicians and physicists have made strong efforts to solve these problems, but for most realistic applications, it was simply a hopeless task. This state of affairs changed dramatically in the middle of the twentieth century. The electronic computer was invented.

In this chapter we cover essentially the period 1945–1960. It contained considerable progress concerning two of the most challenging problem areas in computing at the time, namely, the numerical solution of differential equations and the associated problem of solving large systems of algebraic equations.

4.1 The First Electronic Computers

In the history of mankind, war efforts seem to have given rise to new inventions when it comes to technological developments. During World War II this was even more pronounced. Radio communication was new, and that opened a path for developing and analyzing new methods. Great Britain made strong efforts to break the German coding of their military messages, and for this purpose Alan Turing

---

This chapter contains references to Wikimedia. Wikimedia Commons is a media file repository making available public domain and freely-licensed educational media content (images, sound and video clips) to all.

© Springer International Publishing AG, part of Springer Nature 2018
B. Gustafsson, *Scientific Computing*, Texts in Computational Science
and Engineering 17, https://doi.org/10.1007/978-3-319-69847-2_4

(1912–1954) was employed at the Government Code and Cypher School, Bletchley Park. Turing designed an electromechanical machine, with which he succeeded in breaking the code quite early during the war. This breakthrough is considered as one of the most important strategic achievements for the allied forces to have ended the war. However, Turing's machine was specially defined for the code breaking problem, and could not be used for other problems. But the British continued their efforts to further develop an electronic computer, and established the Computing Machine Laboratory at Manchester University. Turing and Max Newman were the main architects behind the stored program computer called Manchester Small-Scale Experimental Machine (SSEM), which became operational in 1948.

Alan Turing   (Science
Photo Library; 11831246)

Turing had a tragic end to his life. He was a homosexual, and in 1952 he was prosecuted for homosexual acts, which were a crime in Great Britain at that time. In order to avoid prison he accepted chemical treatment for his sexual disposition, but he died in 1954 from cyanide poisoning. Officially it was determined to be a suicide, but it could also have been accidental poisoning. In 2009 Prime Minister Brown made an official apology for the way Turing had been treated by the authorities, and in 2013 Queen Elizabeth II granted him a posthumous pardon.

In the United States at this time, the most urgent task was the Manhattan project aimed at constructing a nuclear bomb. These, and other problems that could be described in mathematical terms, were a driving force that led to the construction of the electromechanical computer, the Automatic Sequence Controlled Calculator (ASCC), also called Mark I. It was built by IBM and the first machine was installed at Harvard University (1944), with the military having priority access.

The most brilliant physicists and mathematicians worked on the Manhattan project, and John von Neumann (1903–1957) was one of them.

John von Neumann   (Alan Richards photographer. From the Shelby White and Leon Levy Archives Center, Institute for Advanced Study, Princeton, NJ, USA)

Von Neumann was an extraordinary man. Born of Jewish parents in Budapest, he was soon found to be a "Wunderkind". At this time children in Hungary didn't begin school until 10 years of age, but von Neumann was taught by governesses much earlier. Not only could he solve advanced mathematical problems as a child, but he also had an extraordinary memory. He could word-for-word recite an article that he had read only once. This extraordinary talent made it possible for him to begin his formal schooling at the famous Fasori Lutheran Secondary School in Budapest, attended by many well known scientists, several of them being Jewish. His father wanted him to become a chemist, so he studied chemistry at the University of Berlin. However, he got his Ph.D. in mathematics at Pázmány Péter University in Budapest at the age of 23, which he achieved at the same time as his studies at the prestigious ETH in Zürich. The next year he started teaching at the University of Berlin. It would have been hard to find anyone who would have had such a solid education and extensive research experience at the age of 25.

In 1930 he moved to the United States and from 1933 to his death he was a professor at the Institute for Advanced Study in Princeton.

Von Neumann had enormous breadth in his research, given his widely different kinds of applications. He was in the forefront of pure mathematics, applied mathematics, physics and engineering with new revolutionary ideas everywhere, and with a keen interest in applications of widely different types. For scientific computing he may be the most influential researcher of all time. Unfortunately, he died way too early of bone cancer at the age of 53.

Von Neumann was heavily involved in the particular problem of designing the implosion technique for initiating a nuclear explosion. The two halves of the uranium charge were to be pushed together by firing a surrounding regular explosive charge. Shortly after the computer Mark I was installed, von Neumann ran a simulation program for the implosion process. Less than a year later, the technique turned out to be successful during the first real bomb test in the New Mexico desert. A few months later, it also worked when it was used for real in Japan.

The Mark I was impressive, but was nothing compared to the new electronic ENIAC machine that was sponsored by the US Army. Its construction started in secret in 1943 at the University of Pennsylvania, and was announced in 1946. The name is interpreted as Electronic Numerical Integrator And Computer, and it paved the way for computations of a magnitude that never had been imagined before. The machine had 17,468 vacuum tubes and a large number of resistors and capacitors. It operated on decimal numbers and had a *program* that broke down the algorithm into a series of simple machine instructions. The program had to be introduced manually by setting the proper connections and switches in a certain way. Data were read into the machine via punch cards, and results were taken out the same way. The introduction of computers based on vacuum tubes meant a revolution in computing power. However, the number of vacuum tubes was very large as we have seen, and they were not very reliable. Immense maintenance was required; on average, 2000 vacuum tubes had to be replaced every month.

The ENIAC computer   (Photo: U.S. Army Photo, public domain. Source: http://.art.mil/ftp/ historic-computers/)

The new machine had a computing speed, the order of magnitude larger than anything seen before. Two numbers could be added in 0.2 ms, but a multiplication of two 10 digit numbers took 2.8 ms. ENIAC and its followers opened up a complete revolution in science. Until then science had the two branches of theory and experiments; now a third branch was born: scientific computing.

At that time von Neumann had become involved in the next Los Alamos project which was the construction of a hydrogen bomb. One of the first test computations on the ENIAC was in fact his program for simulating a key process concerning the bomb. But in addition to becoming a major user of the new computer, he also became involved in the continued development of new computers. He is actually one of the central names when it comes to constructing electronic computers and their programming.

The ENIAC was based on pure electronic components which was a fundamental new invention. The follow up was called EDVAC (Electronic Discrete Variable Automatic Computer), and it was based on an equally fundamental new invention. Vacuum tubes can be either on or off, which means that computations are organized in binary mode. But this makes it natural to store numbers in binary form represented by zeros and ones, nothing else. As an example we have

$$19.25 = 1 \cdot 2^4 + 0 \cdot 2^3 + 0 \cdot 2^2 + 1 \cdot 2^1 + 1 \cdot 2^0 + 0 \cdot 2^{-1} + 1 \cdot 2^{-2} = 10011.01_2,$$

where the last representation is used in the computer, usually with 32 *bits*, in which each one has two states representing either zero or one. EDVAC was the first computer using this binary form, and all later general computers use the same principle. Rules for addition and multiplication of zeros and ones are extremely simple, but as usual one gets nothing without paying in some other way. In this case the price is that a complete addition or multiplication of two general numbers requires many more operations than in the decimal system. But speed is the strength of the electronic circuits, the switch time of the vacuum tube is in microseconds.

But not only did EDVAC use binary number representation as a fundamental new property, it also had a stored program together with the data in its memory. The program was read into the machine via punch cards, and the results came out this way as well. The need for constructing computer programs introduced the *programmer* as a new type of profession that the world had not yet seen.

John Mauchly and Presper Eckert were the main inventors of ENIAC, and they proposed the EDVAC architecture as well. They were joined by von Neumann in a consulting role, and in 1945 he actually wrote the "First Draft of a Report on the EDVAC" as the first official report on the proposed construction.

The American computer EDVAC was completed in 1949, and then several countries started their own development of new computers. Among them was Sweden, where the BESK was introduced in 1953. The name is an acronym referring to the Swedish translation of "Binary Electronic Sequential Calculator". It had 40-bit word length and could do 17,857 additions per second, which actually made it the fastest computer in the world for a short time.

The book *The Computer from Pascal to von Neumann* by H.H. Goldstine, published by Princeton University Press in 1972, describes in detail the development of the first electronic computers and the work of von Neumann.

Applied mathematicians worldwide now became aware of the immense computing power the new computers represented. Some people thought that with this new machine, sometimes called an electronic brain, there was no need to develop any new numerical methods or more advanced types of computers. In fact, there were even those who thought that all necessary computations could be carried out within a few months, and that would be the end of computational mathematics.

What actually happened was quite different. Numerical analysis became a dynamic and fast developing branch of science, and new types of methods have been invented ever since. In contrast to the prewar situation, mathematicians working in this area became much more specialized in numerical analysis. Likewise, specialized computer scientists constructed new types of computer architectures, with an ever increasing speed and storage capacity. The establishment of scientific computing as a third scientific discipline complementing theory and experiment had begun.

4.2   Further Developments of Difference Methods for ODE

As we have seen, finite difference methods for solving ordinary differential equations were well established when the electronic computers entered the field. They were easy to implement, and very early they made a significant impact on computational mathematics with applications in physics, chemistry and engineering. One challenging problem arose with the space age, and in particular the manned missions to the moon. NASA was responsible for the project, and they had already started numerical computations in the 1950s.

Computing the trajectory of the space craft is governed by Newton's laws with variable gravity, and the resulting system of ordinary differential equations is solved as an initial value problem. Stability is of course necessary and high accuracy is required over long time intervals. However, the traditional type of difference methods as described in Sect. 3.7 could be used, and the success of the missions was proof of the reliability of these numerical methods.

However, even if general classes of difference methods were available for many different applications, theoretical questions of practical importance remained and caught the interest of many applied mathematicians.

4.2.1   Stability and Convergence

The ODE numerical communities soon found that stability and accuracy were crucial in understanding finite difference methods as applied to initial value problems, and various observations were made. In 1952 the Swiss mathematician Heinz Rutishauser (1918–1970) published the article [144], in which he gave examples of unstable methods. Furthermore he considered general differential equations and general multistep difference methods (3.20). The corresponding characteristic equation for the homogeneous part is

$$\kappa^k + \alpha_0 \kappa^{k-1} + \cdots + \alpha_{k-1} = 0 \tag{4.1}$$

with roots $\kappa_1, \kappa_2, \ldots, \kappa_k$. Rutishauser gave the fundamental necessary condition for stability:

$$|\kappa_j| \leq 1, \qquad j = 1, 2, \ldots, k.$$

Many other mathematicians started looking into the theory of ordinary differential equations and finite difference methods. One of the most influential in the postwar era was the Swede Germund Dahlquist (1925–2005).

Germund Dahlquist
(Photographer unknown.
Royal Institute of
Technology, School of
Computer Science and
Communication)

He was born in the university town of Uppsala, but started his academic education at Stockholm University at the age of 17. In the Swedish university system there is a pre-doctoral degree called *licentiat* requiring a thesis, and he completed that in 1949. However, instead of continuing for a Ph.D., he started working for the newly formed Swedish Board for Computing Machinery, where he went to work as a programmer on the home-built computer BESK mentioned above. After a few years he changed his mind about a Ph.D. and got his degree in 1959 with his thesis on numerical methods for ODE. For the rest of his long career he worked on this topic, and became one of the world's leading researchers. In 1963, he became the first professor of Numerical Analysis in Sweden. The chair was at the Royal Institute of Technology (KTH), and he kept it until his retirement.

Together with Åke Björck in Linköping, Dahlquist wrote several textbooks in numerical analysis. He was active in founding the journal BIT, and he served as its editor for many years. He also worked for Amnesty International, where he made essential contributions when it came to helping scientists in political trouble.

Dahlquist started developing a general theory for initial value problems, in particular multistep difference methods for first order differential equations. He gave precise definitions for accuracy, stability and convergence, and in 1956 he gave the fundamental theorem for convergence as the stepsize h tends to zero, see [32]. If f is Lipschitz continuous. the approximation v converges to the solution u if and only if the roots of the characteristic polynomial (4.1) satisfy

1. $|\kappa_j| \le 1, \qquad j = 1, 2, \ldots, k.$
2.   The roots located on the unit circle are simple.

4.2.2   Stiff ODE

Dahlquist continued working on important stability problems, and in particular he worked on theoretical questions concerning methods for stiff ODE. He introduced the concept of A-stability, which is of fundamental importance for such differential equations. We will first give a short characterization of such problems.

At the University of Wisconsin, Madison, Charles Francis Curtiss and Joseph Oakland Hirschfelder worked in the chemistry department. The problem is that the formation of the radicals happens on a much faster time scale than the ordinary chemical reaction. Curtiss and Hirschfelder considered the differential equation

$$\frac{du}{dt} = a(t, u)u + F(t, u),$$

where $|a(t, u)|$ is large and $F(t, u)$ behaves as a smooth function of t. In order to understand the difficulty, let us consider the trivial initial value problem

$$\frac{du}{dt} = au + b,$$

$$u(0) = u_0,$$

where a and b are constants with $a \ll 0$. The solution is

$$u(t) = e^{at}u_0 + \frac{b}{a}(e^{at} - 1),$$

and since the exponentials die out very quickly, we get $u(t) \approx -b/a$ after a short time, and we assume that this is the solution of interest. Such a differential equation involving widely different time scales is called *stiff*.

Now consider the Euler method

$$v_{n+1} = v_n + ahv_n + bh,$$

which after n steps has the solution

$$v_n = (1 + ah)^n u_0 + bh \sum_{j=0}^{n-1}(1 + ah)^j = (1 + ah)^n u_0 - \frac{1 - (1 + ah)^n}{a}b.$$

Clearly we need quite a small stepsize with $h < 2/|a|$, such that the amplification factor $1 + ah$ is smaller than 1 in magnitude. The closer the timestep is to $1/|a|$, the faster the convergence is to the steady state solution $-b/a$.

If, instead of the Euler method, we use the backward version

$$v_{n+1} = v_n + ahv_{n+1} + bh,$$

the amplification factor $1 + ah$ is replaced by $(1 - ah)^{-1}$, and we get

$$v_n = (1-ah)^{-n}u_0 + bh \sum_{j=0}^{n-1}(1-ah)^{-j} = (1-ah)^{-n}u_0 - \frac{(1 - (1 - ah)^{-n})(1 - ah)}{a}b.$$

For a fixed timestep h, the solution approaches $-b/a + bh$ for increasing n, and unlike the explicit case, the choice of h is governed only by accuracy considerations.

This was realized by Curtiss and Hirschfelder for the much more complicated differential equation they were dealing with, and they introduced and analyzed the Euler backwards method, see [31]. Furthermore, they discussed second and third order implicit methods, where the approximation of du/dt at $t = t_n$ uses t_n, t_{n-1}, t_{n-2} and t_n, t_{n-1}, t_{n-2}, t_{n-3} respectively. These methods were later called *backward differentiation methods*.

Before describing the theory, we define stiffness somewhat differently from the example above for scalar ODE, and consider instead systems

$$\frac{du}{dt} = Au,$$

where $u = u(t)$ is a vector with m components and A is an $m \times m$ matrix. If the eigenvalues of A are denoted by λ_j, $j = 1, 2, \ldots, m$, the solutions consist of components $e^{\lambda_j t}$, and if the eigenvalues have very different magnitude, these components vary on different time scales. In such a case the system is called *stiff* in analogy with the scalar example above. Such systems are very common in many applications, for example in chemistry, electric circuits, and fluid dynamics. If one is interested only in the fastest varying component, there is not much of a choice when computing the solution. The stepsize h has to be chosen such that the solution is properly resolved, and the choice is essentially between explicit methods. However, the initial vector may be such that the fast varying components are never activated, and the process is essentially governed by eigenvalues of smaller magnitude. This is in analogy with the weather prediction problem discussed in Sect. 3.8.4. This problem is governed by a system of PDEs, but the principle is the same: there are inherent fast varying gravity waves, but these have little influence on the behavior of the weather.

In such a case, the timestep can be chosen to be larger from an accuracy point of view. The fast components are not present and don't need to be represented by a fine grid. However, the important point is that small perturbations in the computational process may be severely enlarged with time if the numerical method is not constructed properly. This is what happened in the example above with the explicit Euler method. The result becomes useless if h is not chosen to be very small.

Dahlquist took a general view of this problem and introduced the concept of *A-stability* defined as follows. Consider the scalar initial value problem

$$\frac{du}{dt} = \lambda u,$$

$$u(0) = 1,$$

where λ is a complex constant with $Re\,\lambda < 0$. The solution is $u(t) = e^{\lambda t}$, and obviously, $\lim_{t\to\infty} u(t) = 0$. If the solution of a certain numerical method exhibits the same asymptotic behavior for all timesteps h, then the method is said to be A-stable.

Consider now a general multistep method (3.20), which for the model equation becomes

$$v_{n+1} + \sum_{j=0}^{k-1} \alpha_j v_{n-j} = z \sum_{j=-1}^{k-1} \beta_j v_{n-j},$$

where $z = \lambda h$. The characteristic equation is

$$(1 - z\beta_{-1})\xi^k + \sum_{j=0}^{k-1}(\alpha_j - z\beta_j)\xi^{k-j-1} = 0,$$

and the method is A-stable if and only if the roots ξ_j satisfy

$$|\xi_j| < 1 \quad \text{for all } z \text{ with } Re\,z < 0.$$

Dahlquist proved that an A-stable multistep method can be at most of second order accuracy; this result goes under the name *the Dahlquist barrier*, see [33]. Furthermore, the most accurate method is the trapezoidal rule

$$v_{n+1} - v_n = \frac{h}{2}\big(f(t_{n+1}, v_{n+1}) + f(t_n, v_n)\big).$$

Stiff systems can be modified such that some of the differential equations are replaced with a set of algebraic equations, resulting in *differential–algebraic equations*, which requires new methods.

We shall come back to later developments of difference methods for initial value problems for ODEs that took place during the later decades of the twentieth century.

4.3 PDE and Difference Methods

We have seen in Sect. 3.8 that a few attempts were made to solve PDE problems in the early twentieth century. The introduction of electronic computers opened up new possibilities to solve such problems numerically, which earlier had been impossible because of the huge manual computations that were involved. The approach for a numerical solution was almost exclusively based on finite difference methods at this time. However, not only were new methods needed for certain applications, but there were also central theoretical questions that remained. For example, how can one guarantee that a certain numerical solution of a difference scheme really is close to the true solution of the differential equation? Furthermore, how can one be sure that the solution on a certain grid becomes more accurate if the grid is refined? The concept of stability and convergence were introduced. The challenge is to construct algebraic criteria that not only guarantee stability, but also are easy to check.

4.3.1 Fourier Analysis and the von Neumann Condition

When using explicit difference methods for time-dependent problems, there is always a limit on the timestep caused by stability requirements. This limit was given for the standard second-order approximation of the wave equation by the CFL-condition discussed in Sect. 3.8.4. But the question is how to find this limit theoretically for a more complicated problem. For the time-dependent PDE, v_j^n is a customary notation for approximating $u(x, t)$ at the grid-point ($x = j\Delta x, t = n\Delta t$). With v^n denoting the long vector with components v_j^n, the difference scheme can be written in the form

$$v^{n+1} = Qv^n, \tag{4.2}$$

where Q can be seen as a matrix with unbounded size as the stepsize Δx becomes arbitrarily small. Stability is then defined by requiring that

$$\|Q^n\| \leq K \tag{4.3}$$

holds for all positive integers n, where the constant K is independent of Δx and n. For hyperbolic problems, the difference operator Q is typically a function of

$\lambda = \Delta t / \Delta x$, but for general problems, it is a very difficult problem to find the limit value of λ such that (4.3) is satisfied. This is true even if the stability condition is strengthened to the stricter but simpler condition

$$\|Q\| \le K.$$

The first published attempt to simplify this analysis was done by John Crank (1916–2006) and Phyllis Nicolson (1917–1968). (A woman in mathematics on this level was quite unusual at this time.) Crank and Nicolson were interested in the numerical solution of the heat equation (3.28), just as Richardson was, as described in Sect. 3.8.3. The simplest difference scheme is

$$v_j^{n+1} = v_j^n + \lambda(v_{j+1}^n - 2v_j^n + v_{j-1}^n),$$

where $\lambda = \Delta t / \Delta x^2$, and for parabolic problems such as this one, it is natural to keep λ as a constant. The solution can be expanded into a sin-series in space:

$$v^n = \sum_\omega \hat{v}_\omega^n \sin(\omega \pi x). \tag{4.4}$$

The relation between the coefficients

$$\hat{v}^{n+1} = \hat{Q}\hat{v}^n \tag{4.5}$$

now becomes very simple, and stability requires

$$|\hat{Q}| \le 1.$$

Basic trigonometric formulas gives the simple relation

$$\sin(\pi\omega x_{j-1}) - 2\sin(\pi\omega x_j) + \sin(\pi\omega x_{j+1}) = -4(\sin^2 \frac{\xi}{2})\sin(\pi\omega x_j),$$

where $\xi = \pi\omega\Delta x$. Hence

$$\hat{Q} = 1 - 4\lambda \sin^2 \frac{\xi}{2}, \qquad |\xi| \le \pi,$$

giving the stability condition $\lambda \le 1/2$, or

$$\Delta t \le \frac{\Delta x^2}{2}. \tag{4.6}$$

The condition (4.6) is a severe restriction on the timestep, in particular the corresponding condition for problems in several space dimensions. Crank and Nicolson therefore suggested the implicit method (3.30), which had been used already by Richardson as a method for computing the first step in a two-step method. The method requires the solution of a tri-diagonal system for each timestep. A sin-series expansion used as above gives

$$|\hat{Q}| = \frac{|1 - 2\lambda \sin^2(\xi/2)|}{|1 + 2\lambda \sin^2(\xi/2)|} \leq 1,$$

i.e., the method is unconditionally stable. Since it is second order accurate both in time and space, one can choose Δt of the same order as Δx.

By using the same type of technique, Crank and Nicolson also showed that the leap-frog scheme (3.29) used by Richardson is unstable. The discrete solution v_j^n can also in this case be expanded into a sin-series (4.4), and the coefficients of this series satisfy a recursion formula in time

$$\hat{v}_\omega^{n+1} + 4\lambda(\sin^2 \xi)\hat{v}_\omega^n - \hat{v}_\omega^{n-1} = 0,$$

where $\lambda = 2\Delta t/\Delta x^2$. The solution of this difference equation has the form

$$\hat{v}^n = \alpha_1 \kappa_1^n + \alpha_2 \kappa_2^n,$$

where κ_1, κ_2 are the two roots of the characteristic equation

$$\kappa^2 + 4\lambda(\sin^2 \xi)\kappa - 1 = 0.$$

We are now in the same situation as for ODE in Sect. 4.2, and also here we need $|\kappa_j| \leq 1$, $j = 1, 2$, otherwise perturbations will grow without bound since n is arbitrarily large. In our case there is always one root with $|\kappa_2| > 1$ for $\lambda > 0$, and consequently the original difference method cannot be expected to work well. It is unconditionally unstable.

The technique used here is a special case of Fourier analysis, which was first used for stability analysis of difference methods by von Neumann. Indeed, the authors acknowledge him in the article by writing that the *"way of examining the accumulation of errors ... was proposed to the authors by Prof. D.R. Hartree, following a suggestion by Prof. J. von Neumann"*.

In the examples above a sin-series could be used as a representation of the solution. If the boundary conditions for a certain problem are substituted by the assumption of 2π-periodic solutions, then one can expand the solution in a discrete Fourier series in space, using the general discrete Fourier transform as discussed in Sect. 3.6. If $N\Delta x = 2\pi$, the grid-function takes the form

$$v_j^n = \frac{1}{\sqrt{2\pi}} \sum_{\omega=0}^{N-1} \hat{v}_\omega^n e^{i\omega x_j}, \qquad j = 0, 1, \ldots, N - 1,$$

and the difference scheme in space/time becomes a difference scheme in time. In the case above, we again get (4.5) with the same $\hat{Q}$ and exactly the same stability conditions.

It is somewhat surprising that this general technique was not published until 1950 in the article [21] with von Neumann as coauthor. This article contains the first weather prediction method that really could be used for practical cases. Charney and his coauthors had already noted that the original set of prediction PDEs used by Richardson could be used if the timestep is chosen small enough such that the fast gravity waves are resolved, see Sect. 3.8.4. However, they note that a prediction over 24 h would then require nearly 100 time cycles which was considered *"a formidable number for the machine that was available to the writers"*. (The machine was the first electronic computer ENIAC.)

It is interesting to note that this article of such importance for the meteorological society also contained a general method for stability analysis of difference methods for time-dependent PDE, leading to the *von Neumann condition*, which by far is the most widely used technique over time until now. It is in general not a sufficient condition for stability, but is often the only condition that can be verified for many applications.

The authors formulate the problem of weather prediction in terms of the vorticity which leads to a scalar PDE in two space dimensions. In this way the difficulty with gravity waves is avoided. Here we use a simpler model equation to illustrate stability analysis. We consider the nonlinear PDE in one space dimension

$$\frac{\partial u}{\partial t} = a(u)\frac{\partial u}{\partial x},$$

and the leap-frog approximation

$$v_j^{n+1} - v_j^{n-1} = \frac{\Delta t}{\Delta x}a(v_j^n)(v_{j+1}^n - v_{j-1}^n)$$

that was used in [21]. The standard simplification technique of linearization is used here, such that the function $a(v_j^n)$ is substituted by a known function $a(x_j, t_n)$. Then we go one step further and make this function constant, such that we get the difference scheme

$$v_j^{n+1} - v_j^{n-1} = \lambda(v_{j+1}^n - v_{j-1}^n), \tag{4.7}$$

where $\lambda = a\Delta t/\Delta x$, a constant.

When the difference operator in space acts on the Fourier component $e^{i\omega x_j}$, we get

$$e^{i\omega x_{j+1}} - e^{i\omega x_{j-1}} = 2\lambda i \sin(\omega \Delta x)e^{i\omega x_j}.$$

This is again the magic of the Fourier technique. In analogy with the effect on differential operators, a complicated difference operator becomes a simple multiplication by a complex scalar. As a consequence, the partial difference equation (4.7) transfers into an ordinary difference equation for the Fourier coefficients

$$\hat{v}_\omega^{n+1} - 2i\lambda \sin(\omega\Delta x)\hat{v}_\omega^n - \hat{v}_\omega^{n-1} = 0.$$

The roots of the characteristic equation

$$\kappa^2 - 2i\lambda \sin(\omega\Delta x)\kappa - 1 = 0$$

are

$$\kappa_{1,2} = i\lambda \sin(\omega\Delta x) \pm \sqrt{1 - \lambda^2 \sin^2(\omega\Delta x)} \tag{4.8}$$

with $|\kappa_{1,2}| = 1$ if $|\lambda| \leq 1$. To be precise, the case $|\lambda| = 1$ must be eliminated, since otherwise there are multiple roots, and the general form of the solution

$$\hat{v}_\omega^n = \alpha_1 \kappa_1^n + \alpha_2 \kappa_2^n$$

does not hold. The final stability condition is therefore $\lambda = a\Delta t/\Delta x < 1$.

In order to implement it for the original nonlinear equation, we must know something about the solution itself. Even so, there is of course no guarantee that the solution stays stable. The solution is not periodic, and the linearization is quite a significant simplification. Indeed the authors are quite cautious about the relevance of the condition (labeled (29) in their article). They write *"Because of the extreme approximative character of the derivation of (29) this stability criterion can be regarded only as a rough directive in the selection of Δx and Δt"*. Furthermore *"The actual values used for Δx and Δt were chosen on the basis of a combination of the above principles and general physical considerations and were ultimately justified when the computation was found to be stable"*.

This was written in 1950, and is probably almost exactly how computations of time-dependent PDEs with finite difference based methods are still handled 67 years later.

Every difference scheme can be formulated in a one-step form. For the leap-frog scheme above, we introduce the vector

$$\mathbf{v}^n = \begin{bmatrix} v^n \\ v^{n-1} \end{bmatrix},$$

and the difference operator

$$\Delta_0 v_j = v_{j+1} - v_{j-1},$$

and write the scheme as

$$\mathbf{v}^{n+1} = Q\mathbf{v}^n,$$

where

$$Q = \begin{bmatrix} \lambda \Delta_0 & I \\ I & 0 \end{bmatrix}.$$

After Fourier transformation we get

$$\hat{\mathbf{v}}^{n+1} = \hat{Q}\hat{\mathbf{v}}^n, \tag{4.9}$$

where

$$\hat{Q} = \begin{bmatrix} 2i\lambda \sin(\omega \Delta x) & 1 \\ 1 & 0 \end{bmatrix}.$$

The eigenvalues of this matrix are the κ-values in (4.8).

For a general k-step scheme, $\hat{Q}$ in (4.9) is a $(k \times k)$-matrix. The von Neumann condition is usually defined as:

The eigenvalues κ_j of the matrix $\hat{Q}$ satisfy

$$|\kappa_j| \leq 1, \qquad j = 1, 2, \ldots, k,$$
$$|\kappa_j| < 1, \qquad \text{if } \kappa_j \text{ is a multiple eigenvalue.} \tag{4.10}$$

This is exactly the condition given at the end of Sect. 4.2 for ODE difference methods.

For systems of p differential equations, $\hat{Q}$ is a $(pk \times pk)$-matrix. The von Neumann condition is then the same, except that there might be multiple eigenvalues κ_j that don't require the stricter form $|\kappa_j| < 1$.

The transformed equation (4.9) is so much easier to analyze compared to the original equation (4.2), and this is the reason why it is the standard way, and sometimes the only way, to analyze stability.

The true necessary and sufficient stability condition for a problem with constant coefficients and periodic solutions is

$$\|\hat{Q}^n\| \leq K, \tag{4.11}$$

where $\| \cdot \|$ is a matrix norm and K is a constant independent of Δx, Δt, ω. This is a nontrivial condition to verify, and we come back to it later.

The computation that was presented in the article described above was a huge advancement in the efforts to predict the weather. It took 24 h to carry out a 24 h prediction, and as the authors write, they *"were just able to keep pace with the weather"*. On the other hand, these 24 h included a lot of manual operations "*namely by the reading, printing, reproducing, sorting, and interfiling of punch cards*" (100,000 punch cards were produced). Furthermore, the ENIAC machine was not easy to program. In the acknowledgement of the article, the authors thank von Neumann's wife Klara for instructions in coding and for checking the final code.

4.3.2 Implicit Methods and Operator Splitting

It seems like parabolic time-dependent problems were of prime interest during the first postwar years. The most frequent model problem in two space dimensions was the simplest form of the heat equation on the unit square:

$$\frac{\partial u}{\partial t} = \frac{\partial^2 u}{\partial x^2} + \frac{\partial^2 u}{\partial y^2}, \qquad 0 \le x \le 1, \ 0 \le y \le 1,$$

$$u(x, 0, t) = g_{x,0}(x), \quad u(x, 1, t) = g_{x,1}(x), \tag{4.12}$$

$$u(0, y, t) = g_{0,y}(y), \quad u(1, y, t) = g_{1,y}(y),$$

$$u(x, y, 0) = f(x, y).$$

This particular simple PDE offers the possibility of expressing the solution in terms of series expansions that can be evaluated explicitly. However, for slightly more complicated equations, for example with variable coefficients $a(x, y)$, $b(x, y)$ multiplying the derivatives, this is no longer possible, and difference methods was the only alternative. However, with the first computers, the problem (4.12) was a real challenge. The methods discussed earlier in this book for problems in one space dimension can easily be generalized to several space dimensions, but the computations become too heavy. We have already discussed the severe limit on the timestep for explicit difference schemes, since this requires that the timestep Δt be chosen proportional to the square of the space-steps Δx, Δy. A remedy for this was introduced by Crank and Nicolson as we have seen in Sects. 3.8.3 and 4.3.1. They considered the one-dimensional case, but it can be generalized in a straightforward way to the two-dimensional problem. With $v^n_{j,k}$ denoting the approximation of $u(x_j, y_k, t_n)$ and the difference operators

$$\delta_x^2 v^n_{j,k} = v^n_{j+1,k} - 2v^n_{j,k} + v^n_{j-1,k},$$

$$\delta_y^2 v^n_{j,k} = v^n_{j,k+1} - 2v^n_{j,k} + v^n_{j,k-1},$$

the Crank–Nicolson method is

$$v^{n+1}_{j,k} - v^n_{j,k} = \frac{\lambda_x}{2}\delta_x^2(v^{n+1}_{j,k} + v^n_{j,k}) + \frac{\lambda_y}{2}\delta_y^2(v^{n+1}_{j,k} + v^n_{j,k}),$$

where $\lambda_x = \Delta t / \Delta x^2$, $\lambda_y = \Delta t / \Delta y^2$. It is unconditionally stable, but there is a severe drawback. There is a system of algebraic equations to be solved for the unknowns $v_{j,k}^{n+1}$ at all inner points of the domain. At this time there were no special direct methods for this type of systems, and consequently direct elimination methods were out of the question. Iterative methods existed, but the convergence properties were uncertain, in particular for more general problems with variable coefficients in the PDEs.

In the beginning of the 1950s, a few high technology companies got their own computers and began using mathematical models and numerical methods for simulating key processes in their development of new technologies. Humble Oil and Refinery Company in Houston, Texas, decided to employ two chemical engineers Donald Peaceman and Henry Rachford who had their ScD-degrees from MIT. Furthermore they employed the mathematician Jim Douglas Jr. who got his Ph.D. at Rice University. Probably the directors did not have any idea about the impact these three people would have on computational mathematics. The problem at hand was how to simulate fluid flow through porous media in order to get a better understanding of how to get the oil out of the ground in an efficient way. In order to avoid the difficulty with solving the huge algebraic system arising from the Cranck–Nicolson method, Peaceman and Rachford came up with the idea of splitting the difference operator into two different parts, each one associated with only one coordinate direction and applying them alternatively on the latest time-level. On an $N_x \times N_y$-grid, the difference scheme is

$$v_{j,k}^{n+1/2} - v_{j,k}^n = \frac{1}{2}(\lambda_x \delta_x^2 v_{j,k}^n + \lambda_y \delta_y^2 v_{j,k}^{n+1/2}), \qquad j = 1, 2, \ldots, N_x - 1,$$

$$k = 1, 2, \ldots, N_y - 1,$$

$$v_{j,k}^{n+1} - v_{j,k}^{n+1/2} = \frac{1}{2}(\lambda_x \delta_x^2 v_{j,k}^{n+1} + \lambda_y \delta_y^2 v_{j,k}^{n+1/2}), \qquad k = 1, 2, \ldots, N_y - 1,$$

$$j = 1, 2, \ldots, N_x - 1.$$

$$(4.13)$$

The new procedure was called the *Alternating-Direction Implicit Method (ADI)*. When Dirichlet boundary conditions are used, the resulting systems have a tri-diagonal structure, and these can be solved by direct elimination. With N_x and N_y of the order N, the solution on each coordinate line requires $\mathcal{O}(N)$ arithmetic operations. Consequently, the total work is $\mathcal{O}(N^2)$ for each full timestep, and this is the same as for a simple explicit difference method.

The authors used Fourier analysis to prove that the ADI-method is unconditionally stable. But Douglas went further. He pointed out that the limit process as the stepsizes Δx, Δy, Δt tend to zero had not been studied in detail, and no convergence proofs were available for any implicit methods. Therefore he gave a complete proof of convergence not only for the new ADI-scheme, but also for earlier developed schemes. For the ADI-scheme and any constant $\lambda_x = \lambda_y = \Delta t / \Delta x^2$ he proved that the error in the numerical solution is of order Δt, or equivalently, of order Δx^2.

The results were presented in two separate articles, but they were published simultaneously in the same issue of *J. Soc. Indust. Appl. Math.*, see [131] and [42].

In many applications, the steady state of the solution is of prime interest. For the model problem, it leads to the Poisson equation

$$\frac{\partial^2 u}{\partial x^2} + \frac{\partial^2 u}{\partial y^2} = F(x, y), \qquad 0 \le x \le 1, \ 0 \le y \le 1,$$

$$u(x, 0) = g_{x,0}(x), \quad u(x, 1) = g_{x,1}(x),$$

$$u(0, y) = g_{0,y}(y), \quad u(1, y) = g_{1,y}(y),$$

where $F(x, y)$ represents a heat source. Peaceman and Rachford advocate the use of their ADI-method for the time-dependent problem as an iterative method for this problem. The timestep is used as a parameter to be chosen for best possible convergence speed. This method was also published, see [131].

The original convergence proof by Douglas was based on an error estimate of order Δt. For the steady state computations the size of the timestep is of course no issue other than the effect on the rate of convergence as $n \to \infty$ with fixed Δt. Indeed, one doesn't even need consistency in time. However, if the true time development is needed, it would be nice to have second-order accuracy in both time and space. After all, first-order accuracy in time is obtained already for the basic explicit method which is much simpler to implement. This problem was of course realized by the authors, and the following year Douglas and Rachford published a more refined error analysis which resulted in an $\mathcal{O}(\Delta t^2 + \Delta x^2 + \Delta y^2)$ error estimate, see [43]. In order to achieve this, the algorithm was formulated differently, and other authors came up with still other different forms. A compact and elegant form is obtained by starting from (4.13) and then subtracting the second equation from the first one, which gives

$$2v^{n+1/2} = v^{n+1} + v^n + \lambda_y \delta_y^2 (v^n - v^{n+1}),$$

where the subscripts have been omitted. When this is substituted into the sum of the two equations, after rearranging we get

$$(I - \lambda_x \delta_x^2)(I - \lambda_y \delta_y^2)v^{n+1} = (I + \lambda_x \delta_x^2)(I + \lambda_y \delta_y^2)v^n.$$

This is an example of a general *operator splitting* technique. In general the original differential operator is partitioned into a sum of d operators, where each operator has a simpler structure. This splitting is then used to define an approximation of the form

$$P_1 P_2 \ldots P_d v^{n+1} = Q_1 Q_2 \ldots Q_d v^n,$$

which is solved as a sequence of equations. This technique is quite common in various types of applications. The operators P_j and Q_j need not be associated with coordinate directions as in the ADI-case. Another typical case is where each operator represents a particular physical process modelled by a differential operator of a certain order that may be different for different operators.

4.3.3   Stability and Convergence

As for ODE, basic questions about stability and accuracy for difference approximations of time-dependent PDEs were in focus in the 1950s. As we have seen, the von Neumann analysis based on Fourier transforms was the standard technique. It is easy to apply and in some cases it is also sufficient for stability. In the previous section we saw that convergence, as the stepsize tends to zero, was proven in a special analysis for implicit schemes that were considered. The challenge at this time was to derive a more general theorem establishing conditions for convergence for general difference methods.

Peter Lax and Robert Richtmyer had both worked at Los Alamos with the Manhattan project; now they were at New York University (for more about Lax, see Sect. 4.3.5). They worked out a beautiful theorem that sometimes is called the fundamental theorem of Numerical Analysis. It deals with pure initial value problems, i.e., either problems on infinite domains or problems with periodic solutions. For simplicity we limit ourselves here to one-step schemes

$$v^{n+1} = Q(\Delta t)v^n,$$

where it is assumed that there is some constant relation between the stepsize in space and the stepsize Δt in time. We want to solve the problem on an interval $0 \le t \le T$. There are four fundamental concepts that must be strictly defined.

- The original initial value problem for the PDE is *well posed* if

$$\|u(t)\| \le C\|u_0\|,$$

  where C is independent of the initial function u_0.
- The difference scheme is *consistent* if for smooth solutions $u(t)$ of the PDE

$$\left\|\frac{u(t + \Delta t) - Qu(t)}{\Delta t}\right\| \to 0 \quad \text{as } \Delta t \to 0.$$

- The difference scheme is *stable* if

$$\|Q^n\| \le K, \qquad n\Delta t \le T,$$

  where K is independent of Δt.

- The difference scheme is *convergent* if

$$\|v^n - u(t_n)\| \to 0 \quad \text{as } \Delta t \to 0.$$

The *equivalence theorem* presented in [112] in 1956 is then simple to formulate:
Assume that the initial value problem for the PDE is well posed. Then a consistent difference approximation is convergent if and only if it is stable.

There are two remarkable facts. (1) There is no way to get convergence with an unstable scheme. (2) There is no assumption on smoothness for the solution u, only that there is a solution with bounded norm (technically we are working in a Banach space).

We also note that this theorem was published in the same year as Dahlquist published his fundamental results for ODE, see Sect. 4.2.

The theorem holds exactly as stated also for multistep methods; we just have to rewrite them in one-step form as described earlier for ODE.

Clearly this theorem makes the analysis of a given method so much simpler. We need only to check consistency and stability, where the first one is trivial. Stability is a little bit more complicated, but as we shall see, many conditions were derived in the following years in order to also make this analysis easy.

Lax and Richtmyer are two of the most influential researchers when it comes to difference methods for PDE.

4.3.4   The Matrix Theorem

We have earlier described the work on ODE problems by Dahlquist. For PDE problems, Stockholm came to be a place where significant progress happened also for PDE, and in particular for time-dependent PDE. Heinz-Otto Kreiss (1930–2015) started working at the Swedish Board for Computing Machinery, where Dahlquist already worked.

Finite difference methods was the dominating class of numerical methods also for partial differential equations at this time. The only practical stability criterion available then was the von Neumann condition as described in Sect. 4.3. However, this condition is necessary but not sufficient for stability. If the difference scheme is written in the form

$$\hat{v}^{n+1} = \hat{Q}\,\hat{v}^n$$

after Fourier transformation, stability can be expressed as

$$\|\hat{Q}^n\| \le K, \qquad n\Delta t \le T, \tag{4.14}$$

where the constant K is independent of n. Kreiss proved the famous matrix theorem which provides three different criteria that are equivalent to (4.14), see [98]. The

Heinz-Otto Kreiss   (Photo:
Anders Sköllermo)

stepsize is a parameter in $\hat{Q}$, in the general case there may be several parameters. The theorem is formulated for a family $\mathscr{F}$ of matrices A, and the fundamental difficulty is that the estimates must be uniform for all matrices in the family.

The theorem is as follows:

Consider all matrices $A \in \mathscr{F}$. The following conditions are equivalent:

- There is a constant K such that $\|A^n\| \le K$.
- There is a constant K such that $\|(\lambda I - A)^{-1}\| \le K/(|\lambda| - 1)$ for all complex scalars λ with $|\lambda| > 1$ (*the resolvent condition*).
- There is a constant K, such that for every A, there is a nonsingular matrix S with $\|S\| \le K$, $\|S^{-1}\| \le K$, such that $\tilde{A} = SAS^{-1}$ is upper triangular with its elements satisfying

$$|\tilde{a}_{jj}| \le 1 \quad \text{for } 1 \le j \le m,$$

$$|\tilde{a}_{jk}| \le K \min(1 - |\tilde{a}_{jj}|, \, 1 - |\tilde{a}_{kk}|) \quad \text{for } j < k.$$

- There is a constant K, such that for every A, there is a matrix H such that

$$K^{-1}I \le H \le KI \quad \text{and} \quad A^*HA \le H.$$

The matrix theorem is great mathematics, but Kreiss himself admits in the paper that the stability conditions are "*not easy to apply*". We shall see in Sect. 4.3.6 how he found other conditions that were easier to apply.

Kreiss had quite an interesting life story. He was born in Hamburg 1930, a very gifted child. However, because of the war and the massive bombing of the city he had to move to the countryside as a young teenager, where he worked for farmers in order to get enough food and firewood for the family he stayed with. After the war he moved back to Hamburg, where he completed his education up to a basic degree at Hamburg University. In 1955 he moved to Stockholm, where he worked for the meteorological institute at the university and ran his programs on the new Swedish computer BESK at the newly formed computer center. In 1960 he got his Ph.D. in mathematics at the Royal Institute of Technology (KTH) with his thesis [97] written in German; in English translation the title was "On the solution of initial-value problems for partial differential equations".

After a year as professor at Chalmers Institute of Technology in Gothenburg, he accepted a newly installed professor chair in Numerical Analysis at Uppsala University in 1965. There he attracted good students, and a strong group was created which within a decade came to be well known internationally. He was a frequent visitor at American universities, and 1978 he left Uppsala for a new professorship at the California Institute of Technology (Caltech) in Pasadena, California. Later he switched to UCLA in Los Angeles, where he worked until he returned to Stockholm in 2003.

Kreiss had an enormous impact on the development and analysis of numerical methods for time-dependent PDE. He was also a very good advisor, and had a large number of Ph.D. students. Today there are several hundred researchers, engineers and academic teachers who have had Kreiss or his students or their students as advisors.

4.3.5   Fluid Dynamics and Shocks

The Manhattan project at Los Alamos had very high priority in the US, and many of the sharpest scientists were involved. One of the questions was how the detonation wave associated with a nuclear explosion behaves. Fluid dynamics (including gas dynamics) is governed by a set of nonlinear partial differential equations that were first developed by Euler. In one space dimension they are

$$\frac{\partial \rho}{\partial t} + \frac{\partial (\rho u)}{\partial x} = 0,$$

$$\frac{\partial (\rho u)}{\partial t} + \frac{\partial (\rho u^2 + p)}{\partial x} = 0, \tag{4.15}$$

$$\frac{\partial E}{\partial t} + \frac{\partial \left(u(E + p) \right)}{\partial x} = 0.$$

Here ρ, u, p, E denote density, particle velocity, pressure and energy respectively. Nonlinearity makes it impossible to find any solutions by analytical means, except for trivial cases of little interest. We have described Richardson's work in Sects. 3.8.3 and 3.8.4, but not much more had been done on this topic.

Von Neumann was heavily involved in the Manhattan project, and again, as for so many other problems in mathematics, physics and engineering, he was the one who made the first serious attempts to solve problems in gas dynamics using numerical methods. He started by considering the problem of gas flow in a tube with a closed end. If the gas is pushed towards the end, a discontinuity in the pressure and particle velocity is formed, and this creates a fundamental computational difficulty.

In order to illustrate basic difficulties with nonlinear PDEs, we shall consider a simple example with *Burgers' equation*

$$\frac{\partial u}{\partial t} + u\frac{\partial u}{\partial x} = 0 \,,$$

in the interval $-1 \le x \le 1$ with the boundary condition $u(-1, t) = 1$, and an initial function as shown in Fig. 4.1. As long as the solution has some degree of smoothness, the solution moves with the speed u, which is the coefficient of $\partial u/\partial x$ in the differential equation. Accordingly, the upper part of the solution is moving faster than the lower part, and at $t = 1$ a discontinuity will be formed.

Now there is a big problem. At $x = 0$, the solution jumps from $u = 1$ to $u = 0$, and there is no longer any unique speed of propagation. We would expect a continued movement of the shock, but at what speed?

We shall come back to this question, but we note first that von Neumann tried to solve the shock tube problem described above by applying a difference method directly to the differential equations. Almost everything at Los Alamos was classified, but in 1944 he wrote a report on his findings, see [157]. He showed that the numerical solution had a strongly oscillating character near the shock and was useless for any conclusions about the true physical behavior.

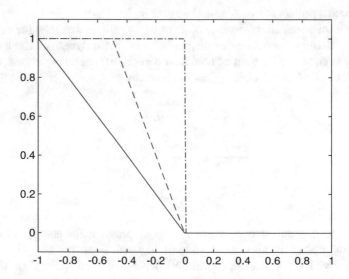

Fig. 4.1 Solution of Burgers' equation, $t = 0$ (dash-dotted line), $t = 0.5$ (dashed line), $t = 1$ (solid line)

Von Neumann was frustrated about the slow progress in fluid dynamics. In a talk in Montreal in 1945 he indicated that the only way towards a better understanding was to develop new computers and numerical methods. He wrote (see [111]):

We could, of course, continue to mention still other examples to justify our contention that many branches of both pure and applied mathematics are in great need of computing instruments to break the present stalemate created by the failure of the purely analytical approach to nonlinear problems. Instead we conclude by remarking that really efficient high-speed computing devices may, in the field of nonlinear partial differential equations as well as in many other fields which are now difficult or entirely denied of access, provide us with those heuristic hints which are needed in all parts of mathematics for genuine progress. In the specific case of fluid dynamics these hints have not been forthcoming for the last two generations from the pure intuition of mathematicians, although a great deal of first-class mathematical effort has been expended in attempts to break the deadlock in that field. To the extent to which such hints arose at all (and that was much less than one might desire), they originated in a type of physical experimentation which is really computing. We can now make computing so much more efficient, fast and flexible that it should be possible to use the new computers to supply the needed heuristic hints. This should ultimately lead to important analytic advances.

Let us now go back to the simplified problem with Burgers' equation above. There is no fluid that is truly inviscid, even if the viscosity is extremely small. Viscosity is modelled by a second derivative in the differential equation, so we introduce such a term with a small coefficient ε:

$$\frac{\partial u}{\partial t} + u\frac{\partial u}{\partial x} = \varepsilon\frac{\partial^2 u}{\partial x^2}.$$

The perturbation is singular in the sense that the character of the PDE has changed from hyperbolic to parabolic. There is a smoothing effect on the solutions, and von Neumann came up with the idea to use that for computational purposes. In 1950 he and Robert Richtmyer wrote the first article with a new idea for using artificial viscosity for numerical computation, see [158]. In this way they didn't have to keep track of the position of the shock and thereby they avoided any special operations in a neighborhood. In their article they write *"Then the differential equations (more accurately, the corresponding difference equations) may be used for the entire calculation, just as though there were no shocks at all"*. These types of methods were later called *shock capturing methods*. They have a coefficient ε that is decreasing with a finer grid.

Figure 4.2 shows the result for our problem with $\varepsilon = 0.01$ when using a standard second order difference method. Parabolic equations have smoother solutions, and we can see that the shock now has rounded shoulders. But we have a unique solution, and the shock moves with the speed $u = 0.5$, which is the average of the solution before and after the shock. When using further decreasing values of ε, the shock becomes sharper and sharper.

Besides von Neumann, another Hungarian and extremely bright mathematician joined the Los Alamos team as a young student, namely Peter Lax. He was born in Budapest 1926, but being Jewish, his family left Hungary for United States in 1941.

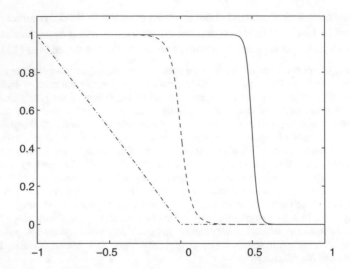

Fig. 4.2 Solution of the viscous Burgers equation, $t = 0$ (dash-dotted line), $t = 1$ (dashed line), $t = 2$ (solid line)

After Los Alamos and courses at a few universities, he went to New York University in 1946 and got his Ph.D. there in 1949 with Richard Courant and Kurt Friedrichs as supervisors and coworkers.

Lax was influenced by the work of von Neumann and Richtmyer, and their introduction of artificial viscosity for shock problems. He continued to work on nonlinear PDEs and shocks for decades. In fact he is probably the leading researcher of all time in this area, combining pure mathematics with the development of

Peter Lax   (IMS Bulletin, Institute of Mathematical Statistics, IMS. https://owpdb.mfo.de. Oberwolfach Photo Collection)

numerical methods. He got the prestigious Abel Prize in 2005 with the motivation
*"for his groundbreaking contributions to the theory and application of partial
differential equations and to the computation of their solutions"*. In line with the
concept of artificial viscosity for difference methods, he actually introduced it for
the pure mathematical theory of shocks. It is not at all trivial as to how to define a
unique solution that makes sense for the real physical world. Lax introduced what
he called weak solutions in his 1954 article [109]. With the notation $u(x, t, \varepsilon)$ for
the viscous solution, he defined the inviscid solution $u(x, t)$ in strictly mathematical
terms as the limit

$$u(x, t) = \lim_{\varepsilon \to 0} u(x, t, \varepsilon).$$

He followed up with another article [110] in 1957, and published several more on
this topic during the following years.

The physically correct value of ε for typical applications is very small and cannot
be used for practical calculations. There is a delicate balance here; the viscosity
coefficient ε must be chosen large enough such that the computations are kept stable,
while at the same time small enough to keep accuracy sufficiently good. In a 1954
article Lax presents what is now called the Lax–Friedrichs scheme. Consider the
general equation

$$\frac{\partial u}{\partial t} + \frac{\partial}{\partial x} f(u) = 0, \tag{4.16}$$

where $f(u)$ is a nonlinear function of u. The numerical method is

$$v_j^{n+1} = \frac{1}{2}(v_{j+1}^n + v_{j-1}^n) - \frac{\Delta t}{2\Delta x}\big(f(v_{j+1}^n) - f(v_{j-1}^n)\big).$$

It seems odd that the more obvious choice v_j^n as the first term at the right-hand side is
replaced by $(v_{j+1}^n + v_{j-1}^n)/2$, but the reason is stability. We consider the simple linear
case with $f(u) = au$, where a is a constant. The corresponding approximation is

$$v_j^{n+1} = \frac{1}{2}(v_{j+1}^n + v_{j-1}^n) - \lambda(v_{j+1}^n - v_{j-1}^n), \qquad \lambda = \frac{a\Delta t}{2\Delta x}.$$

The amplification factor in Fourier space is

$$\hat{Q} = \cos \xi - \lambda i \sin \xi, \qquad |\xi| \leq \pi$$

with $|\hat{Q}| \leq 1$ for $\lambda \leq 1$. Not only do we get a stable scheme, but there is also a
damping of all nonzero frequencies, except for $\xi = 0$ and $\xi = \pi$. However, the last
exception is not good. The corresponding nonphysical oscillations initiated by the
presence of discontinuities will stay without damping.

Later Lax developed another method together with Burt Wendroff, both working at what is now the Courant Institute in New York. The method is based on a Taylor expansion in time, and then using the differential equation for transferring time-derivatives to space-derivatives. For the linear model equation

$$\frac{\partial u}{\partial t} = a \frac{\partial u}{\partial x},$$

we get

$$u(t + \Delta t) = u(t) + \Delta t a \frac{\partial u}{\partial x}(t) + \frac{a^2 \Delta t^2}{2} \frac{\partial^2 u}{\partial x^2}(t).$$

By replacing the derivatives with centered second-order difference approximations, the result is the Lax–Wendroff method

$$v^{n+1} = v^n + \frac{\lambda}{2}(v^n_{j+1} - v^n_{j-1}) + \frac{\lambda^2}{2}(v^n_{j+1} - 2v^n_j + v^n_{j-1}), \qquad \lambda = \frac{a \Delta t}{\Delta x}.$$

It is second order accurate in both time and space, and a Fourier analysis gives

$$\hat{Q} = 1 + \lambda i \sin \xi - 2 \sin^2 \frac{\xi}{2}, \qquad |\xi| \leq \pi;$$

and it then follows easily that the von Neumann condition becomes $\lambda \leq 1$. Furthermore, if $\lambda < 1$, then $|\hat{Q}| < 1$ for all nonzero ξ, i.e., there is damping for all nonzero frequencies. Such a method is called dissipative.

Several other versions have been developed, but the original Lax–Wendroff scheme is by now classic, and serves as the typical representative of the so-called shock capturing methods, where no special operations are implemented at or near the shock.

Sergei Godunov (Picture included with list of visitors to ICASE in 2002. http://www.icase.edu/images/visitor/2002/godunov-sergei.jpg. Creator: ICASE, NASA Langley Research Center, Hampton, Virginia. Wikimedia*)

Fig. 4.3 Shock trace and propagation direction of information

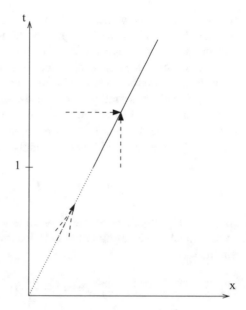

A different approach was used by Sergei Godunov (1929–). Already in his Ph.D. thesis (1954) at the Moscow State University he considered the problem of avoiding oscillatory solutions around discontinuities and gave fundamental results concerning monotone methods that preserve a monotone discrete function from one time level to the next. In 1959 he published his landmark paper [70] on a new method, which came to be called the Godunov method. We shall first explain the idea by going back to the Burgers equation above.

As long as no discontinuity has developed, there is no real difficulty getting any difference scheme to work, but the trouble starts at $t = 1$. Figure 4.3 shows the x/t-plane and the trace of the discontinuity. The dashed arrows indicate how a certain feature is propagated. While the direction of the propagation direction varies continuously before $t = 1$, for $t \geq 1$, there is a discontinuity across the shock. This is the reason for the nonphysical oscillations that we have described above. Godunov took a fundamentally different approach to get rid of the trouble, and here is how it works.

To begin with, the PDE is written in the conservation form (4.16), where for the Burger equation $f(u) = u^2/2$. At $t = t_n$ the numerical solution at the grid-points is u_j^n. Define a function

$$\tilde{u}(x, t) = u_j^n, \qquad x_j - \frac{\Delta x}{2} < x < x_j + \frac{\Delta x}{2}$$

for all j at $t = t_n$, i.e., $\tilde{u}$ is a piecewise constant function with discontinuities in the middle of every interval. A *Riemann problem* for a certain conservation law (4.16) is the problem of finding the solution when the initial function, which is equal to a constant a everywhere to the left of a certain point x_0 and equal to another constant b everywhere to the right. For the Burgers equation, we have already seen that this solution propagates the discontinuity with speed $(a + b)/2$ and keeps the same constant values on each side. Godunov generalized the Riemann problem to an initial function with several discontinuities. Since there is a finite speed of propagation, there is a unique solution also to this problem provided the time interval is small enough, because then the features originating at a half-point $x_{j+1/2}$ do not collide with the features from its neighbors. Therefore we choose t_{n+1} such that $\tilde{u}(x, t_{n+1})$ is well defined for all x, and then take the average

$$u_j^{n+1} = \frac{1}{\Delta x} \int_{x_j - \Delta x/2}^{x_j + \Delta x/2} \tilde{u}(x, t_{n+1}) dx$$

to get the grid-values at the new time-level. For the example above, with a monotonically decreasing initial function, the Godunov method reduces to the very simple form

$$u_j^{n+1} = u_j^n - \frac{\Delta t}{\Delta x} \left((u_j^n)^2 - (u_{j-1}^n)^2 \right).$$

This is called an *upwind scheme*, since it takes its values from the side from where the relevant information for the true differential equation is coming. That principle was already used in 1952 by Courant, Isaacson and Rees, see [29]. However, Godunov's method is more general. He presented his method for the Euler equations (4.15), which is a system that also can be written in a conservation form

$$\frac{\partial \mathbf{u}}{\partial t} + \frac{\partial}{\partial x} \mathbf{f}(\mathbf{u}) = 0$$

with

$$\mathbf{u} = \begin{bmatrix} \rho \\ \rho u \\ E \end{bmatrix}, \qquad \mathbf{f}(\mathbf{u}) = \begin{bmatrix} \rho u \\ \rho u^2 + p \\ u(E + p) \end{bmatrix}.$$

The system is first integrated in space to obtain

$$\frac{d}{dt} \int_{x_{j-1/2}}^{x_{j+1/2}} \mathbf{u}(x, t) dx = \mathbf{f}\big(\mathbf{u}(x_{j-1/2}, t)\big) - \mathbf{f}\big(\mathbf{u}(x_{j+1/2}, t)\big).$$

This is an exact formula, and it is the basis for *finite volume methods* that are constructed by devising various ways of approximating $\mathbf{u}$ and integration in time. Godunov's idea was to represent $\mathbf{u}$ as a piecewise constant vector function and then integrate it exactly in time, i.e., solve the Riemann problem at every discontinuity. These solutions were well known also for the Euler equations, and a completely new method for simulating inviscid flow had been invented.

Seldom is there a new numerical method with such a different structure compared to earlier ones as was the invention by Godunov. When the general trend was to introduce artificial viscosity directly into the numerical approximation in order to remove discontinuities, Godunov introduced discontinuities (!) at every grid-point. Furthermore he used analytical solution methods as an intermediate step when defining the numerical solution at the new time level. His method is an example of a truly remarkable achievement based on a very original way of thinking.

Godunov's work did not get much attention in the west until a few decades later. But then the idea became the basis for many types of generalizations; and so-called Riemann solvers were developed for more complicated problems, sometimes in the form of approximate solvers. In this way, accuracy of the original method had been raised to a higher degree, and we now have a collection of *Godunov type methods*.

When applied to shock problems, all methods we have described so far are of the shock capturing type. Actually, even the Godunov method has a viscosity effect built in, which makes discontinuous solutions look like the viscous solution in Fig. 4.2. However, there are also *shock fitting methods* that leave shocks as true discontinuities. These are constructed such that the location of the shock is kept as part of the solution, and the numerical solution is then found separately on each side.

The conditions provide algebraic conditions relating the values of the dependent variables on one side of the shock to the values of the corresponding variables on the other side of the shock. In this way they serve as local boundary conditions at the shock. Aerodynamics is a typical application where shock fitting has been frequently used, in particular for solving problems of the same type as the blunt body problem. When a body travels fast in the air, a shock forms in front of it, with the air unaffected in front of the shock. This makes the Rankine–Hugoniot conditions easy to apply, and the computation can be limited to the area between the shock and the body.

A different approach to handling discontinuities was used by Stanley Osher and James Sethian when they introduced the *level set method* in 1988, see [128]. It turns out that the same principle can be used for many other problems, and it has became rather popular in various applications. Instead of describing it here for the application to fluid dynamics, we introduce it later in Sect. 5.15 for its application to image processing.

4.3.6 Stability for Dissipative Methods

Stability for approximations of parabolic PDEs is relatively easy to verify. The reason is that there is an inherent damping in the PDE itself, which is most easily seen in Fourier space. For the equation $\partial u/\partial t = \partial^2 u/\partial x^2$, the Fourier coefficients satisfy

$$\hat{u}_\omega(t) = e^{-\omega^2 t}\hat{u}_\omega(0).$$

The corresponding Fourier amplification factor $\hat{Q}$ for a difference scheme is typically inside the unit circle, at least for Δt small enough, and finding the limit for Δt (if there is any) becomes easy. Hyperbolic PDE is more of a challenge. For the simplest possible hyperbolic equation $\partial u/\partial t = \partial u/\partial x$, the Fourier coefficients are

$$\hat{u}_\omega(t) = e^{i\omega t}\hat{u}_\omega(0),$$

which shows that there is no damping at all. Here we typically get amplification factors for the difference methods that should be located close to the unit circle, and the stability might be a much more delicate matter.

Hyperbolic PDEs arise naturally in many applications. Various wave propagation problems like acoustics and electromagnetics lead to such equations, but also inviscid flow problems. In general, for all mathematical models that describe energy conserving processes we can expect difficulties for numerical methods, since the perturbation introduced by the discretization may go the wrong way. In the previous section we demonstrated how artificial viscosity was introduced for nonlinear shock problems. This was the driving force towards the more general concept of dissipative numerical methods, even for linear problems.

Kreiss, who had proved the famous matrix theorem, now turned to the problem of finding general methods and stability conditions that were easy to verify, and he found it natural to use the dissipativity concept with the Lax–Wendroff scheme as a typical model. He considered general hyperbolic systems of PDE

$$\frac{\partial u}{\partial t} = A(x)\frac{\partial u}{\partial x},$$

where A is a symmetric matrix. The difference scheme is

$$v_j^{n+1} = Q(x_j, \Delta x)v_j^n, \tag{4.17}$$

where it is assumed that there is a relation $\Delta t = \lambda \Delta x$, where λ is a constant. A straightforward Fourier transformation does not lead anywhere, since $Q(x_j, \Delta x)$ requires a Fourier expansion as well as v_j^n, and there will be a multiplication of the two. Consequently, there is no simple relation between the Fourier coefficients. However, for any x, one can formally define the matrix $\hat{Q}(x, \xi)$ with $\xi = \omega \Delta x$, where ω is the wave number and $|\xi| \leq \pi$. The symbol $\hat{Q}(x, \xi)$ cannot be used for computing the solution in Fourier space, but it can be used for stability analysis of the original scheme. The approximation is defined as dissipative of order $2r$ if the eigenvalues κ_j of $\hat{Q}$ satisfy

$$|\kappa_j| \leq 1 - \delta|\xi|^{2r}, \qquad \delta > 0$$

for every fixed x. The theorem presented in [99] in 1964, (actually stated for several space dimensions) then says that the difference scheme is stable if it is dissipative of order $2r$ and accurate of order $2r - 1$. Later this condition was extended such that it covers also the case where the order of accuracy is $2r - 2$ such as the Lax–Wendroff scheme with $r = 2$.

The concept of dissipativity came to be important when developing new robust difference methods, and Kreiss' theory made it easy to analyze the stability properties.

4.4   Progress in Linear Algebra

With the introduction of electronic computers came the possibility to solve very large linear systems of algebraic equations, not the least were those that occurred as a result of discretizing differential equations. Eigenvalue problems were another challenge, also those often arising as a consequence of applying numerical methods to differential equations. Iterative methods attracted most of the interest, but the size of the systems and the large number of iterations meant that convergence properties became important. Various methods and tools for analysis were developed starting right after the war, and in this section we shall describe some of the more important work in this area.

4.4.1   Orthogonalization

All computational techniques must be such that the results are reasonably resistant against perturbations in the data. Vectors and functions are often expressed in terms of certain basis vectors and functions, and then orthogonality plays an important role. We shall illustrate this with a very simple example.

Consider a two-dimensional vector space, where two non-orthogonal vectors are used as a basis, with the coordinates denoted by ξ and η, see Fig. 4.4.

We want to compute the distance r to the point with the known coordinates (ξ_0, η_0). We do this by applying the Pythagorean theorem on the right angle triangle with sides η_0, b and r. We have

$$a = \frac{\xi_0}{\cos \theta},$$

$$b = \frac{\eta_0 - a}{\tan \theta},$$

$$r = \sqrt{a^2 + b^2}.$$

Assume now that the angle θ is small, i.e., the two basis vectors are almost linearly dependent. An error ε in η_0 results in an error $\varepsilon / \tan \theta$ in b, which causes an error of the same order in r.

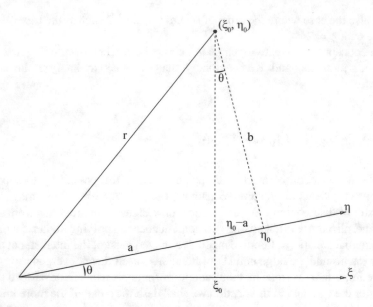

Fig. 4.4 Vector space with non-orthogonal basis vectors

As an example, we get

$$(\xi_0, \, \eta_o) = (1.00, \, 1.01) \quad \Rightarrow \quad r = 1.15.$$

When perturbing the η-coordinate by 1%, we get

$$(\xi_0, \, \eta_o) = (1.00, \, 1.01 \times 1.01) \quad \Rightarrow \quad r = 1.52,$$

which is a 32% error.

The problem with almost linearly dependent vectors arises in many different applications. One way of transforming such a set is the Gram–Schmidt orthogonalization process named after the Danish mathematician Jørgen Pedersen Gram (1850–1916) and the German mathematician Erhard Schmidt (1876–1959). However, it had been introduced earlier by both Laplace and Cauchy. The original version is actually unstable, but was later modified so that it becomes stable. A good description of the full history is given in [114].

However, other methods for orthogonalization were developed later. We shall describe them as applied for the least square problem, which is of great importance in many different applications. We consider it here for overdetermined linear systems of equations

$$A\mathbf{x} = \mathbf{b},$$

where A is an $m \times n$ matrix, $m > n$. Since $\mathbf{x}$ has only n elements, it cannot satisfy the system with m equations in the general case. But we can compute the least square approximation as described in Sects. 3.4 and 3.5. The unique solution of the problem is given by the normal equations

$$A^T A \mathbf{x} = A^T \mathbf{b}.$$

Unfortunately, this system is often quite ill-conditioned, leading to poor results if standard methods are applied. The reason is that the columns of the matrix A are almost linearly dependent, and just as for the trivial example above, we are in for trouble. The cure is orthogonalization.

The nowadays established method is based on QR factorization. Assume that we can factorize A such that

$$A = QR,$$

where R is an upper triangular $n \times n$ matrix. The $m \times n$ matrix Q has orthonormal columns q_j, i.e., they are not only orthogonal, they also have length 1, i.e., $\|q_j\| = 1$. Since $Q^T Q = I$, we now have

$$R^T Q^T \mathbf{b} = A^T A \mathbf{x} = R^T Q^T Q R \mathbf{x} = R^T R \mathbf{x}.$$

This is of course the same normal equations as before, but written in a form that is much more convenient. Unlike the original form, we now have a square leading matrix R^T that has an inverse and can be factored out. This gives the system

$$R\mathbf{x} = Q^T \mathbf{b},$$

where the matrix R is upper triangular. This means that once the factorization is computed, the solution is easily obtained with $\mathscr{O}(m \times n)$ arithmetic operations.

The next question is how to compute the two matrices Q and R. This problem was solved by Alston Scott Householder (1904–1993) and James Wallace Givens (1910–1993) independently of each other and using different methods in the 1950s. Given any vector $\mathbf{u}$ of length 1, the Householder transformation matrix is defined by

$$H_{\mathbf{u}} = I - 2\mathbf{u}\mathbf{u}^T,$$

which obviously is orthogonal and symmetric. Denote by $\mathbf{a}_j$ the columns of A, and the unit vectors by $\mathbf{e}_j$. To obtain the upper triangular matrix R, we begin by zeroing out all elements of the first column except for the first one. This is achieved by choosing

$$\mathbf{u} = \mathbf{u}_1 = \frac{\mathbf{a}_1 - \|\mathbf{a}_1\|\mathbf{e}_1}{\|\mathbf{a}_1 - \|\mathbf{a}_1\|\mathbf{e}_1\|}.$$

The next choice of $\mathbf{u}_2$-vector is made by taking the vector $H_{\mathbf{u}_1}\mathbf{a}_2$ and replacing the first element with zero, thus guaranteeing that the first column is not changed. Denoting this vector by $\mathbf{b}_2$, we have

$$\mathbf{u}_2 = \frac{\mathbf{b}_2 - \|\mathbf{b}_2\|\mathbf{e}_2}{\|\mathbf{b}_2 - \|\mathbf{b}_2\|\mathbf{e}_2\|}.$$

The new matrix has only zeros in its first two columns below the diagonal. This procedure is now continued, such that in the end

$$R = H_{\mathbf{u}_n} H_{\mathbf{u}_{n-1}} \dots H_{\mathbf{u}_1} A.$$

From the relation $A = QR$, we then get by using orthogonality and symmetry

$$Q = H_{\mathbf{u}_1} H_{\mathbf{u}_2} \dots H_{\mathbf{u}_n}.$$

The Givens method for QR decomposition has a similar structure in the sense that Q and R are constructed in a sequential manner. However, the matrices $H_{\mathbf{u}_j}$ are now substituted by Givens rotations.

In the first step the matrix has the form

$$G_{1,m} = \begin{bmatrix} 1 & & & & \\ & \ddots & & & \\ & & 1 & & \\ & & & c & -s \\ & & & s & c \end{bmatrix},$$

where $c = \cos\theta$, $s = \sin\theta$. The computation can be explained by a simple 2×2 example where the second element of a certain vector $[a\ b]^T$ is to be zeroed out such that

$$\begin{bmatrix} c & -s \\ s & c \end{bmatrix} \begin{bmatrix} a \\ b \end{bmatrix} = \begin{bmatrix} r \\ 0 \end{bmatrix}.$$

The angle θ is not calculated explicitly; the formulas are

$$r = \sqrt{a^2 + b^2},$$
$$c = a/r,$$
$$s = -b/r.$$

The first transformation annihilates the lower left element of A, followed by the annihilation of the next to last element of the new first column using the matrix

$G_{1,m-1}$, etc. The transformation

$$G_1 A = G_{1,2} G_{1,3} \cdots G_{1,m} A$$

now produces a matrix, where the first column has zeroes everywhere, except in the first position. Then the second column is treated in the same way, and finally we obtain an upper triangular matrix.

4.4.2 Iterative Methods for Linear Systems of Equations

The idea of iteration when solving certain problems is very old; we have seen in Sect. 2.2 how Archimedes used it for computing π. Early attempts to use it for solving linear systems of algebraic equations were made by Jacobi (mentioned in Sect. 3.1.2). He used it for systems

$$A\mathbf{x} = \mathbf{b}$$

by splitting the matrix as $A = D + B$, where D is the diagonal of A. The iteration formula is

$$\mathbf{x}^{(n+1)} = -D^{-1} B\mathbf{x}^{(n)} + D^{-1}\mathbf{b}. \tag{4.18}$$

If the matrix is *diagonally dominant*, i.e.,

$$|a_{jj}| > \sum_{k \neq j} |a_{jk}|, \qquad j = 1, 2, \ldots, N,$$

then the method converges to the true solution, and the convergence is faster the more the diagonal elements dominate.

We have seen in Sect. 3.5 that already Gauss used iteration in using a method that was later published by von Seidel. This method is very similar to the Jacobi method, the difference being that the latest available computed values are used, i.e., $x_k^{(n)}$ on the right-hand side of (4.18) is substituted by $x_k^{(n+1)}$, $k = 1, 2, \ldots, j - 1$ in equation no. j. The matrix B with the diagonal excluded is now split into a lower triangular part L and an upper triangular part U such that $A = L + D + U$. The Gauss–Seidel method can then be written in the form

$$(L + D)\mathbf{x}^{(n+1)} = \mathbf{b} - U\mathbf{x}^{(n)}.$$

Even if the formula looks more complicated than the Jacobi method, it does not require any more effort to solve. By starting from the top, it is simply a division of known values by the diagonal element a_{jj} for each equation. The convergence conditions are less strict, but still quite demanding. And fast convergence, which is required for large systems, may be impossible to obtain.

This was the state of affairs with the introduction of electronic computers and its boost to PDE-solvers with their accompanying large algebraic systems that must be solved. Better methods were needed.

Garrett Birkhoff was a professor at Harvard University, when he got David Young as a new Ph.D. student in 1947. Young's thesis was completed in 1950, and it came to be one of the quite rare examples which has an immediate impact on the associated research area. At the 50-year anniversary, Gene Golub and David Kincaid wrote in the preface of a republished version of the thesis: *"David Young's thesis is one of the monumental works of modern numerical analysis. His creation, development and analysis of the Successive Overrelaxation (SOR) method has been fundamental in our understanding of iterative methods for solving linear equations"*. So, what was this SOR method about?

With the notation above, the original system can be written in the form

$$(D + \omega L)\mathbf{x} = \omega\mathbf{b} - \big(\omega U + (\omega - 1)D\big)\mathbf{x},$$

where $\omega > 1$ is a scalar parameter. The iteration method is

$$(D + \omega L)\mathbf{x}^{(n+1)} = \omega\mathbf{b} - \big(\omega U + (\omega - 1)D\big)\mathbf{x}^{(n)}.$$

Again, one iteration is no more demanding than the Jacobi method.

The principle for the introduction of the relaxation parameter ω is easier understood if we note that each equation of the Gauss–Seidel method can be expressed in the form

$$x_j^{(n+1)} = x_j^{(n)} + r_j^{(n)}, \tag{4.19}$$

where

$$r_j^{(n)} = \frac{1}{a_{jj}}\Big(b_j - \sum_{k=1}^{j-1} a_{jk}x_k^{(n+1)} - \sum_{k=j}^{n} a_{jk}x_k^{(n)}\Big).$$

The SOR method is then obtained as

$$x_j^{(n+1)} = x_j^{(n)} + \omega r_j^{(n)},$$

and with $\omega > 1$ a larger weight is given to the residual. This is the reason for the label "overrelaxation".

At this time it was natural to consider systems that were derived from difference approximations of elliptic PDE, and the definitions of certain key properties of the matrix arose from these assumptions. Convergence properties could be strictly proven under certain conditions on the matrix that were quite natural for these kinds of problems. The SOR method was the standard iterative method of choice for several decades.

A remarkable fact is that the same principle with overrelaxation was introduced independently by Stanley Frankel at Caltech, Pasadena, California, and published the same year, see [58]. He presented convincing numerical experiments for the superiority of the method when applied to the Laplace equation and the biharmonic equation on a square with Dirichlet boundary conditions.

No matter what numerical method is used for solving a linear system of equations, there is always an error in the solution (except for very exceptional cases). This is due to the fact that all operations are carried out with finite precision on numbers represented by a finite number of digits. The question must then be asked if the error in the solution is of the same order as the error in the given data for the original system. This problem was considered by the British mathematician James H. Wilkinson (1919–1986). During the Second World War he was working on important applied problems such as ballistics, and as part of this work he was forced to carry out many iterations on large linear systems using a hand calculator. After the war he joined the National Physics Laboratory in Teddington, first working with Turing on the computer construction project, but later with problems in linear algebra.

In the effort to obtain error bounds for the solution, Wilkinson introduced a new original idea, namely the backwards error analysis. For a linear system of equations

$$A\mathbf{x} = \mathbf{b},$$

we assume that the solution obtained from the computer is $\tilde{\mathbf{x}}$. A traditional error analysis is asking for an estimate of the type

$$\|\tilde{\mathbf{x}} - \mathbf{x}\| \le \epsilon,$$

where $\| \cdot \|$ denotes some vector norm and ϵ is some number to be found. However, finding such an estimate might be very difficult. The alternative backwards analysis is asking what problem the obtained solution really solves, i.e., what is $\tilde{A}$ and $\tilde{\mathbf{b}}$ in the system

$$\tilde{A}\tilde{\mathbf{x}} = \tilde{\mathbf{b}} \ ?$$

An exact solution can of course not be obtained, but we are looking for an estimate

$$\|\tilde{A} - A\| \le \epsilon_1,$$

$$\|\tilde{\mathbf{b}} - \mathbf{b}\| \le \epsilon_2,$$

where ϵ_1 and ϵ_2 hopefully are small. Wilkinson developed the whole machinery for finding such estimates, also for eigenvalue problems. This new technique was quite revolutionary, and set the standard for error analysis ever since.

4.4.3   *Eigenvalues*

We have used the eigenvalues of a matrix several times above as a method for analysis of stability and convergence. One can generalize this concept to other types of operators. For any kind of linear operator A, the *eigenvalue problem* is

$$Av = \lambda v.$$

Here one is looking for a complex number λ and a nonzero function or vector v belonging to a space where the operator is well defined. For example, if A is the differential operator d/dx and the space consists of functions of one independent variable x, we have

$$\frac{d}{dx}e^{\lambda x} = \lambda e^{\lambda x},$$

i.e., any number λ is an eigenvalue and $e^{\lambda x}$ is an eigenfunction.

From a computational point of view, the most common type of eigenvalue problem is for $m \times m$ square matrices A and vectors v. The matrix A is often a discretization of some operator involving differential operators, and m is usually a very large number. The problem can be written as a system of algebraic equations

$$(A - \lambda I)v = 0.$$

Since we are looking for a nontrivial solution v, the matrix $A - \lambda I$ must be singular, i.e.,

$$\mathrm{Det}(A - \lambda I) = 0.$$

The left-hand side is a polynomial of degree m in λ. For $m \geq 5$, there is no explicit formula for the roots, and it is necessary to use numerical methods. However, it turns out that even the best numerical root-finders are way too slow for realistic values of m. It is necessary to take advantage of the fact that A is a matrix.

The first numerical methods for computing the whole set of eigenvalues were based on the fact that the matrix $S^{-1}AS$ has the same eigenvalues as A. If such a *similarity transformation* produces a diagonal matrix, then the problem is solved, since the eigenvalues are found in the diagonal. But the problem is of course that S is not known. However, as usual in computational mathematics, it is possible to approach the final form by iteration. We have seen earlier in this book that Jacobi was the pioneer in constructing new numerical methods in different fields, and he was also the first to come up with an effective algorithm for computing eigenvalues.

Let $S_{pq}(\phi)$ be a matrix that equals the identity matrix except for the elements

$$s_{pp} = s_{qq} = \cos \phi,$$

$$s_{pq} = -s_{qp} = \sin \phi,$$

i.e., it has the form

$$
S_{pq} = \begin{bmatrix}
1 & & & & & & & & \\
& \ddots & & & & & & & \\
& & 1 & & & & & & \\
& & & c & & s & & & \\
& & & & 1 & & & & \\
& & & & & \ddots & & & \\
& & & & & & 1 & & \\
& & & -s & & c & & & \\
& & & & & 1 & & & \\
& & & & & & & \ddots & \\
& & & & & & & & 1
\end{bmatrix},
$$

and since $S_{pq}^T(\phi)S_{pq}(\phi) = I$, it is orthogonal. This is a *plane rotation*, often called a *Jacobi rotation* after its inventor. Starting out from the original matrix $A_0 = A$, Jacobi's idea was to define new matrices A_j sequentially by similar transformations so that the off-diagonal elements become smaller and smaller in magnitude. His strategy was to find the largest nondiagonal element in magnitude at each step. If this element is found in position (p, q), the new matrix is defined as

$$
A_{j+1} = S_{pq}(-\phi)A_jS_{pq}(\phi),
$$

where the angle ϕ is chosen so that the element a_{pq} of the new matrix is zero. Since

$$
S_{pq}^{-1}(\phi) = S_{pq}^T(\phi) = S_{pq}^T(-\phi),
$$

this is a similar transformation which means that each new matrix has the same eigenvalues as the previous one, and one can hope for fast convergence.

The method was not used much during Jacobi's time due to the heavy work that is required without access to computers. However, it turns out that the algorithm converges quite fast in practice, even if one can find bad cases. Furthermore, it is very robust, since only orthogonal transformations are involved.

The only other pre-computer attempt to find a numerical method for the eigenvalue problem was the *power method*. This was published in [156] in 1929 by the German mathematicians Richard von Mises and Hilda Pollaczek–Geiringer (born in Vienna), but there has been some hints about earlier use of the method. It is a very simple algorithm; with a given initial vector v_0, it is

$$
v_{j+1} = \frac{Av_j}{\|Av_j\|}, \qquad j = 0, 1, \dots.
$$

If the iterations converge, the direction of v_j tends to the direction of the eigenvector corresponding to the largest eigenvalue in magnitude. This eigenvalue is obtained as

$$\lambda_1 = \frac{v_j^T A v_j}{v_j^T v_j}.$$

However, in order to guarantee convergence, it is necessary that this eigenvalue be strictly greater than the others in magnitude.

With computers available in the 1950s, Wallace Givens proposed an eigenvalue algorithm based on Jacobi type rotations, but in the first stage of making the matrix tridiagonal, this can be achieved in a finite number of steps.

Another way of zeroing out "non-tridiagonal" elements was proposed by Householder, and we described the principle in Sect. 4.4.1. The Householder transformation is represented by the Hermitian and orthogonal matrix

$$P = I - 2uu^T,$$

where u is a vector of unit length. The implementation of these transformations is made without forming the Householder matrix explicitly, and it leads to a very effective algorithm.

The Swiss mathematician Heinz Rutishauser (1918–1970) working at ETH, Zürich, was interested in eigenvalue computation. His method could be interpreted as an LU algorithm, i.e., it was based on a factorization $A = LU$, where L and U are lower and upper triangular matrices respectively. This factorization can be achieved by Gaussian elimination, possibly after pivoting, as we have seen. Rutishauser published his algorithm in [145] in 1958, and called it the LR algorithm. It can be applied to any matrix A and is defined by the iterative procedure

$$A_j = L_j R_j,$$
$$A_{j+1} = R_j L_j, \qquad j = 0, 1, \ldots$$

with $A_0 = A$. Here L_j is lower triangular with ones in the diagonal, and R_j is upper triangular. Since

$$A_{j+1} = R_j L_j = L_j^{-1} A_j L_j,$$

the new matrix has the same eigenvalues as the previous one at each step. Under certain conditions, A_j tends to an upper triangular matrix as $j \to \infty$, and the eigenvalues are found in the diagonal.

If some of the eigenvalues are close to each other, the convergence is slow, and if they coincide, a special treatment is necessary. However, the main difficulty is that the algorithm is in general not very stable. We have seen that orthogonal transformations are preferred from a stability point of view for many methods,

and so it is also for this problem. The LR-algorithm was modified accordingly by
the British mathematician and computer scientist John G.F. Francis (1934–), who
published the QR-algorithm 1961–1962, see [56] and [57]. Independently of him,
the Russian mathematician Vera N. Kublanovskaya developed the same algorithm
and published it the same year, see [104].

The structure of the iterations is exactly the same as for the LR-algorithm, but
the L_j-matrix is replaced by an orthogonal matrix Q_j, i.e., $Q_j^{-1} = Q_j^T$:

$$A_j = Q_j R_j,$$

$$A_{j+1} = R_j Q_j, \qquad j = 0, 1, \ldots, .$$

Each matrix R_j is upper triangular, and under certain conditions, the matrices A_j
converge to an upper triangular matrix as well.

There are many implementations of the QR-algorithm, but most of them use
a preprocessor. We have seen above that a symmetric matrix can be taken to
tridiagonal form by a finite number of Givens or Householder transformations. In
the general case, the result of these transformations is an upper *Hessenberg form*,
where all elements below the first subdiagonal are zero, and at this stage the QR-
algorithm is initiated.

A different type of algorithm was invented in 1951 by Walter Arnoldi (1917–
1995), see [3]. It is related to the power method described above, where powers A^j
of the matrix are generated. The space of all vectors

$$\{\mathbf{x}, \, A\mathbf{x}, \ldots, A^N\mathbf{x}\}$$

is called a Krylov space, which we shall discuss further in Sect. 5.11. The Arnoldi
method can be seen as an orthogonalization of these vectors, and the matrix A is
transformed to an upper Hessenberg matrix H_N expressed in the form

$$Q_N^* A Q_N = H_N,$$

where O_N is an orthogonal matrix. The eigenvalues of H_N are then computed by
using for example the QR-method. If A is symmetric, the matrix H_N is tridiagonal,
and then the method is the Lanczos method, which was invented by the Hungarian
mathematician Cornelius Lanczos (1893–1974).

Since the method requires a lot of storage, and also for improving convergence,
the algorithm is often restarted several times; this method is called *Implicitly
restarted Arnoldi method (IRAM)*. Our experience indicates how N should be
chosen before each restart, but theoretically it is still an open problem.

For most problem areas, the development of numerical methods still continues in
the search for higher efficiency. When it comes to eigenvalue problems for general
matrices, it seems that the QR-method, with initial transformation to the Heisenberg
form, is the final answer. Indeed it has been classified by several authors as one
of the ten most important algorithms of the twentieth century. Strangely enough,

Francis didn't realize the impact of his work, which was carried out at the British government agency National Research Development Corporation (NRDC). The reason was that he switched to Computer Science right after the invention of the QR-algorithm. It was not until he was contacted by Gene Golub in 2007 that he became aware of his fame.

Chapter 5
The Establishment of Scientific Computing as a New Discipline

At the end of the 1950s the use of computational mathematics was well established. New electronic computers had been constructed, then, with vacuum tubes replaced by semiconductors, which made them faster and much more reliable. Many problems in science and engineering were solved numerically, and played a role at universities and in certain advanced industries when designing new products. However, many really heavy and difficult problems could still not be solved with sufficiently accurate and reliable results. Weather prediction is a typical example. Even with the availability of sufficiently advanced mathematical models, and even if sufficiently accurate weather data at a given time would have been fed into the computer, the discretization would have had to be fine enough to give reasonable accuracy. At that time any prediction of weather over a 24-h period was quite uncertain.

In this chapter we shall describe the process of developing new computational techniques; we begin with a section on the continuing development of new computer architectures.

5.1 Faster and Faster Computers

The increasing use of scientific computing is based on new numerical methods together with the fast development of computers. During the first decades after the second world war, each new type of computer was designed to run algorithms arising from problems in science and engineering. This development has continued up to the present time, and has resulted in ever faster supercomputers, but during the last

---

*Wikimedia Commons is a media file repository making available public domain and freely-licensed educational media content (images, sound and video clips) to all.

© Springer International Publishing AG, part of Springer Nature 2018
B. Gustafsson, *Scientific Computing*, Texts in Computational Science and Engineering 17, https://doi.org/10.1007/978-3-319-69847-2_5

decades there has also been the development of convenient personal computers that can also be used for scientific computing.

5.1.1 New Hardware

The transistor was invented in 1947, and in the 1950s they replaced the vacuum tubes in computers. Not only were transistors smaller and demanding less energy, they were also much more reliable. At the same time the magnetic-core memory became the standard type of storage. It consisted of a large number of small rings, and the zeros and ones in the binary numbers were represented by the clockwise and counterclockwise direction of the magnetization in them.

The race for faster computers intensified, and it was almost exclusively limited to the United States with IBM and the new company CDC (Control Data Corporation) as the main players. As mentioned in Sect. 4.1, the original fixed point representation of numbers was soon replaced by floating point numbers of the form

$$a \cdot 2^b,$$

b is an integer exponent expressed in binary form. In the scientific computing society, the speed of a computer was measured in *Mflops*, that is, "one million floating point operations per second". One operation means a basic operation like an addition of two numbers. IBM's fastest computer in 1959 was the new IBM 7090, with a speed of 0.1 Mflops; CDC came up with the 3000 series in 1962. Seymour Cray worked for CDC, and he had great plans in aiming for a really fast machine based on new technology. In 1964 the CDC 6600 was finished, and had a speed of 3 Mflops. Since it was much faster than anything else on the market, it was called a *supercomputer*. After continuing with the even faster CDC 7600, Cray decided to start a new company of his own. The first machine called Cray-1 is probably the most well known supercomputer of all time. It was presented in 1976 and was technically very different from previous computers. Besides looking interesting and eye catching, the C-shaped design allowed for short wiring in such a way that the limitation caused by the finite speed of light could be minimized.

The Cray-1 had a special new invention that actually had an impact on the user when it came to developing numerical methods. This was the vector processor. The idea had been introduced earlier in the CDC STAR 100 computer, but was much more refined by Cray. The motivation for a vector processor is that large parts of a typical algorithm consist of a sequence of operations being applied to a data set called vectors. In a traditional processor, the sequence is applied in full on each single data item and then stored before taking on the next item. A vector processor works like an assembly line, i.e., the sequence of operations is applied to the vector so that each operation is applied one-by-one to single data items. In this way each item in the vector is entered into the assembly line as soon as the preceding operation is completed. In order to use the full capacity of such a computer, the algorithm has

Cray-1   (US Dept Of
Energy/Science Photo
Library; 11895862)

to be constructed so that the vector structure dominates. But this is a restriction, since a certain operation on a given element of the vector cannot use the results of the operation on earlier elements. For example, the iterative method

$$x_{n+1} = g(x_n), \qquad n = 0, 1, \ldots,$$

where $g(x)$ is a computable function of the scalar variable x, is no good when it comes to speed on a vector computer compared to a scalar computer. On the other hand, if $\mathbf{x}$ is a vector with N elements and A is an $N \times N$ matrix, the iteration

$$\mathbf{x}^{(n+1)} = A\mathbf{x}^{(n)}, \qquad n = 0, 1, \ldots$$

is very fast if N is large.

The Cray-1 was followed in 1982 by Cray X-MP. It had two vector processors, and therefore was called a *parallel vector computer*. With a proper algorithm allowing for operations simultaneously on different data, the computer was extremely fast with a peak performance of 400 Mflops. Later came the even faster Cray Y-MP and Cray-2, the latter with 1.9 Gflops as top speed, i.e., 1.9 billion operations per second. The Japanese company Fujitsu also came out with high performance vector computers. They all required a lot of energy and gave off a considerable amount of heat. The extremely compact design contributed further to the need for very efficient cooling. Cray-2 had 8 vector processors arranged so that the architecture became very tight. It had liquid cooling with a separate cooling unit. The Cray computers pushed the design of their central processors to the limits in order to increase speed.

But another very different idea emerged. The microprocessor had been invented, and the price decreased of these much simpler processors. The idea of parallel processors had been around for some time, but in the 1970s the first architectures containing a large number of processors were realized. The first one with very high prestanda was built by the Burroughs Corporation and was called ILLIAC-IV. It had 256 processors, and its name was taken from Illinois Automatic Computer alluding

Cray-2 (Source: http://gimp-savvy.com/cgi-bin/img.cgi?ailswE7kkmL1216740; NASA, public domain)

to the University of Illinois at Urbana-Champagne, which had an important role in the project.

The number of processors quickly increased with new machines, and one started talking about *massively parallel computers*. One of the more well known was designed by Danny Hillis at MIT, who founded the Thinking Machines Corporation. They built the Connection Machine which had several thousand processors. One version had 8 hypercubes, in which each one could hold up to 8096 processors.

The concept of supercomputer was later replaced by the High Performance Computer (HPC). It has no real definition, but today the architecture is almost exclusively of the parallel type, and has a much higher capacity than can be obtained by single processor computers. The peak performance of a parallel architecture can of course only be attained if all processors are active simultaneously. This requires that their algorithms and programs be designed so that this is possible. This was a challenge that applied mathematicians didn't have in the old days.

Constructing electronic computers came to be a competitive field as soon as the ENIAC was presented. During the first decades after the war, the United States dominated the development of computers, and it did so also when supercomputers entered the market. But Japan engaged in the competition, and in 1994, Fujitsu broke the world record by introducing the "Numerical wind tunnel" which had 166 vector processors, together providing 282 Gflops. The name was a result of the participation of the National Aerospace Laboratory of Japan, and the machine was mainly used for problems in aerodynamics. Two years later Hitachi presented the Hitachi SR2201 which theoretically could do 614 Gflops.

At the turn of the century still another country entered the HPC-computer arena, and not surprisingly it was China. Actually, their computer Sunway TaihuLight is now (November 2017) at the top when it comes to speed. It was constructed by the (NRCPC), National Research Center of Parallel Computer Engineering & Technology, and has 10.65 million processors, or cores that are nowadays used to denote each unit that operates in parallel. Interestingly, its predecessor Tianhe-2 was based on Intel processors, and there were plans to use Intel technology again. However, since USA claims that export of such products violates a trade embargo on equipment that can be used for developing nuclear weapons, an export ban was imposed. Consequently the Chinese developed their own technology, and they are now independent and do not rely on components from foreign computers.

The proper unit for speed has now become Pflops (10^{15} flops), and the Sunway TaihuLight can do 125 Pflops. Naturally, the largest supercomputers require a lot of power; this one needs 15 MW. It is a significant amount, and corresponds roughly to the power requirement of a medium sized passenger ship at cruising speed.

The development of electronic computers is breathtaking, and it is hard to find any other field with the capacity of such growth. ENIAC didn't use floating point digital representation, but could do 357 additions of two 10-digit numbers per second. This should be compared to Sunway TaihuLight, which is $3.5 \cdot 10^{14}$ times faster.

Until now the performance data we have given refers to the theoretical top speed. However, with the entrance of vector processors and parallel architectures came the problem of utilizing the theoretical maximum performance of the computer. It is very hard to find a realistic problem where this speed can be attained, and therefore the performance when the computer is working on a typical practical problem is more interesting. But what kind of problem should such a benchmark be?

Jack Dongarra has been very active in developing the software package LIN-PACK that contains algorithms for solving various problems in linear algebra. He wanted to know how computers performed when running this system. In the LINPACK users guide [41], he and the father of LINPACK Cleve Moler as the leading authors introduced the benchmark problem in 1979. It provided the solution to a dense linear system of equations with a certain accuracy. Since the number of arithmetic operations is known, the number of flops per second is obtained by timing the calculation. The notation R_{max} is used for this number, while R_{peak} is used for the theoretical peak performance. This is much like rating an automobile when it comes to fuel consumption. The manufacturers optimize their cars for the testruns on a certain driving cycle, which always results in a lower number than the one that can be achieved in practical driving.

Until the early 1990s, R_{peak} was the only parameter given for the speed of computers, but nowadays, R_{max} is more common for measuring performance. It is considerably smaller, and the relation R_{max}/R_{peak} varies of course with different computers. For Tianhe-2, R_{max} is 34 Pflops, which is 61% of R_{peak}. This is quite low, but reflects the difficulties when using a complicated architecture in an optimal way. In general the R_{max} values vary roughly between 60 and 80% of R_{peak} among different computers.

The storage capacity is of course another important parameter when it comes to large problems. ENIAC had a memory that could hold only 20 numbers, and today this would be hard to imagine how it could do any meaningful computations. But by clever programming and efficient algorithms it could. Its memory size increased at a similar rate as its speed, but one must distinguish between random access core memory and various kinds of external memory, which has a much slower access time. The type of such external memories has varied over the years with magnetic tape in the seventies followed by floppy disks and solid state storage. The architecture of the memory has become increasingly complicated, often with more than two levels.

The description of computer development given above is concentrated on the front line machines over the years. Clearly, these machines were not available to everyone who needed computing. Over the years many medium size computers appeared on the market for more everyday use. In the early seventies central systems of computers became available via distributed terminals, permitting many users to get access at the same time, and this is still the most common use of high capacity computers.

However, the most revolutionary development was the introduction of personal computers (PC) with a microprocessor as the central unit. In the beginning only real experts could handle a computer, but already in 1962 John Mauchly wrote: "There is no reason to suppose the average boy or girl cannot be master of a personal computer". His vision became true about a decade later when the first microprocessors occurred, and with them the possibility to build minicomputers. The Xerox Alto was presented in 1973, and it had its own operating system, Furthermore, it was the first to have a graphics display, which was a new fundamental invention that came to revolutionize the future development of PCs. A modified version was put on the market in 1981, and then the market just exploded. The concept of a personal computer was created; they became smaller, and soon they were down to a size so that they could be put on an office desk. The main architect behind the Xerox Alto was Charles Thacker, and in 2009 he received the prestigious Turing Award, which is sometimes considered the Nobel Prize of computing.

In the beginning PCs were used by engineers and university researchers for computing purposes, but with new operating systems they became easier to handle, and so began the introduction to the general market. Together with the development of the internet, PCs became very popular, and today they are quite common in society in general for just about any purpose. When it comes to scientific computing they are certainly well suited as well, and nowadays speed in the Gflops range is not unusual. Many software systems for computing are available, as for example MATLAB (Sect. 5.1.2).

Lately, the concept of *cloud computing* has emerged. The name refers to the sight of clouds in the sky, where there are individual objects partly overlapping, but where no details can be observed. The idea is that many computer systems are connected in a network, and a user has access to it, but doesn't know which computers are used to process the algorithm. It is much like an electric network, in which a user gets electricity, but doesn't know what kind of power stations are used for generating it and where they are located. The advantage is the ease of access, and more importantly, no initial investment has to be made for an increased need of computer power, and one doesn't sit with unused resources when this goes down.

Referring to the extremely fast development of computer hardware with new types of architecture invented all the time, it is safe to say that no one can predict what kind of hardware will be used for scientific computing in a few decades from now.

5.1.2   Programming Computers

The electronic computers introduced possibilities for carrying out computations on a scale that had been unimaginable earlier. But how should one handle the quickly growing demand for programming the numerical algorithms that were needed? The first computers were constructed so that a certain operation or instruction was represented by a specific number. The program consisted of a sequence of such numbers which was entered into the machine, and this was a very cumbersome procedure. This *machine code* was soon replaced by an assembly language, where the instructions were given names that had some bearing on what actually happened in the machine. Furthermore a certain address to the location in the memory was given a symbolic name. The complete assembly program had to be transformed to a machine code program using a software system called *assembler*.

Programming became much more convenient this way, but it didn't take very complicated algorithms to still give very lengthy programs with the risk of introducing a number of bugs. Furthermore, an assembly language was constructed for the specific computer it was intended for, and could not be used for any other type of computer. There was a need for a general type of high level programming language that was general and such that it could be used for many types of computers. The first attempt to create such a language was made by John Backus (1924–2007) and his coworkers at IBM, and the first implementation was made on the IBM 704 computer in 1957. The name of the language was FORTRAN derived from FORmula TRANslating System. It turned out to be a very successful language for scientific computing applications, and various later versions of FORTRAN are actually still used.

FORTRAN was intended to be used on any type of computer designed for scientific computing. This required a *compiler*, which is a program that transforms the *Source code* to a machine code that can be run on the specific computer at hand. This is a challenge by itself. Not only must the machine code be correct, but it must also produce the numerical result in an optimal way as specified by the FORTRAN program. (A skilled programmer can of course produce an even more efficient machine code since he/she doesn't have to take the restrictions imposed by the FORTRAN program into account.)

In the late 1950s an American/European committee came up with a new very general high level language called ALGOL with its name derived from ALGOrithmic Language. John Backus was also instrumental here as a member of the committee. The second version ALGOL 60 published in 1960 became well known, but was used mainly in the academic community. Interestingly, even if it had never become near FORTRAN as a tool for computation, it nevertheless became a way of publishing algorithms in scientific journals for some time.

FORTRAN has been overtaken to a large extent by other languages like C, C++ and Java, even if these are not specialized for application to scientific computing. However, there are many FORTRAN programs still in use, and in a sense this poses a problem. How should these programs be handled and maintained now and in the future if no one knows the language?

There is still another language that really is specialized. It was developed by Cleve Moler at the University of New Mexico at the end of the 1970s, and was given the name MATLAB alluding to MATrix LABoratory. Moler had earlier been active in the development of LINPACK and EISPACK, which are libraries of subroutines for solving various problems in linear algebra, and he wanted to have a programming system that would allow for the convenient use of these programs. MATLAB (since 1984 marketed by MathWorks Inc. cofounded by Moler) can be used either by entering single commands, in which each one results in a number or an array of numbers, or it can be used as a prewritten program executed as a whole. The system quickly became very successful, and it has kept its status over the years as a convenient and much used tool in scientific computing.

More advanced program systems have been developed over the years, for the purpose of being able to solve problems in science and engineering without having to program the whole algorithm. This is a fast developing area, and it is a quite safe prediction that in the future problems in scientific computing will be solved by a simple introduction of the parameters defining the whole problem. The system then would analyze the given problem, choose the proper numerical method and the associated algorithm and finally run it with the given result having an error estimate. Such systems exist already, but will be extended to new applications and further refined.

Finally it should be mentioned that a fundamentally different branch has been developed for scientific computing. It is based on symbolic manipulation, where in contrast to numerical methods, mathematical procedures are carried out exactly. For example, this is possible in some cases for the problem of integrating functions that are given on an explicit form. The problem of finding the integral $\int \cos(x/\alpha)dx$ is easily solved and gives the answer $\alpha \sin(x/\alpha)$. This type of symbolic processing can be very helpful, not least as a complement to numerical algorithms. The first such system was REDUCE developed in the early 1960s by Anthony Hearn, who used the computer language LISP developed by John McCarthy at Stanford University. Today many similar systems are available, some of the most well known being Maple, Mathematica and Maxima. In the latest versions all have quite impressive capabilities and are very useable in the right situations.

5.2 Mathematical Aspects of Nonlinear Problems

For the purpose of making the presentation of numerical analysis easy to understand, we have used simple linear problems as examples. For such problems, the existence of unique solutions is usually possible to verify, while for problems in real life this is not the case, the reason being that the equations are in most cases nonlinear. For this illustration we use an example from algebra, and consider the polynomial equation

$$x^n + a_{n-1}x^{n-1} + a_{n-2}x^{n-2} + \ldots + a_0 = 0, \qquad n \geq 1.$$

For $n = 1$, the equation is linear, and there is a unique solution $x_1 = -a_0$. For $n > 1$, the equation is nonlinear, and there is no longer a unique solution. If multiple roots are counted with their multiplicity, there are exactly n solutions. For $n \leq 4$, they can be expressed explicitly in terms of square roots and higher order roots and there is no need for numerical methods. For $n > 4$ it is well known that there is no such explicit expression in terms of elementary functions, and the roots must be computed numerically. For very special cases, it is possible to solve the equation, for example in the trivial case when the polynomial can be expressed in the factored form

$$a_n(x - x_n)(x - x_{n-1})\ldots(x - x_1) = 0.$$

For nonlinear differential equations the situation is quite similar, but more difficult to resolve. With proper initial and/or boundary conditions, linear differential equations have a unique solution just as algebraic equations. For nonlinear differential equations the existence of a solution has not been proven at all in most cases, much less in a unique solution. In some cases, the existence is verified for certain initial and boundary data but not for the general case.

For nonlinear ordinary differential equations of the form

$$\frac{du}{dt} = F(t, u(t)),$$

$$u(t_0) = f,$$

there is the famous Picard theorem proven by the French mathematician Emile Picard (1856–1941) (son-in-law of Charles Hermite). It can be formulated as:

If $F(t, y)$ is continuous in t and Lipschitz continuous in y, then there is a value $\varepsilon > 0$ such that there is a unique solution in the interval $[t_0 - \varepsilon, t_0 + \varepsilon]$.

So, even if we know that there is a solution, it is not certain that it exists in the whole time interval of interest.

For partial differential equations, the situation is much worse. As an example we take the incompressible Navier–Stokes equations describing incompressible fluid flow:

$$\frac{\partial u}{\partial t} + u\frac{\partial u}{\partial x} + v\frac{\partial u}{\partial y} + w\frac{\partial u}{\partial z} + \frac{\partial p}{\partial x} = v \triangle u + F_1(x, t),$$

$$\frac{\partial v}{\partial t} + u\frac{\partial v}{\partial x} + v\frac{\partial v}{\partial y} + w\frac{\partial v}{\partial z} + \frac{\partial p}{\partial y} = v \triangle v + F_2(x, t),$$

$$\frac{\partial w}{\partial t} + u\frac{\partial w}{\partial x} + v\frac{\partial w}{\partial y} + w\frac{\partial w}{\partial z} + \frac{\partial p}{\partial z} = v \triangle w + F_3(x, t),$$

$$\frac{\partial u}{\partial x} + \frac{\partial v}{\partial y} + \frac{\partial w}{\partial z} = 0.$$

Here

$u, v, w:$ velocity components in the space directions x, y, z respectively,

$p:$ pressure,

$v:$ viscosity coefficient,

$F_j, j = 1, 2, 3:$ external force,

$\triangle = \partial^2/\partial x^2 + \partial^2/\partial y^2 + \partial^2/\partial z^2.$

These equations are of fundamental importance when it comes to studying turbulence. In two space dimensions there are some existence results, but the conditions on the given data are quite strict. In three space dimensions there are no existence results at all. In fact it is given as one of the problems, in which the one who solves it will be awarded the 100,000 USD Millennium Prize.

In France there were many mathematicians who worked on various nonlinear PDE and the existence of solutions. Jacques-Louis Lions (1928–2001) at École Polytechnique was one of them, representing the successful French school of mathematics. Here we have another famous mathematician who attracted talented Ph.D. students who later became quite famous themselves. For example, Roland Glowinski, Jean–Claude Nédélec, Pierre–Arnaud Raviart and Roger Temam were students of Lions, and they all made significant contributions in applied mathematics and numerical methods, in particular finite element methods for PDEs. His son Pierre–Louis Lions (not having his father as a Ph.D. mentor) developed a quite impressive theory for nonlinear PDEs including existence proofs; indeed he received the prestigious Fields Medal in 1994. His books [117] and [116] contain an excellent description of the currently available theory concerning the Navier-Stokes equations.

Jacques–Louis Lions
(Photographer Konrad Jacobs
License: Creative Commons.
Wikimedia*)

Despite the lack of existence proofs, large scale numerical solutions are computed by just about every research and engineering group in fluid dynamics, and the results are certainly useful for many different applications. This is the case for many other nonlinear PDEs, and this is a somewhat strange situation. A significant part of the work in scientific computing is devoted to computing solutions that one doesn't even know whether or not they exist. It is attempting to assert that computational mathematicians are ahead of the applied but more theoretical mathematicians. But of course, the proof of theoretical existence theorems is much more difficult than the construction of numerical methods.

5.3 Refinement of ODE Methods

As we have seen earlier in this book, numerical methods for ordinary differential equations had a quite early start with Euler. Since the work actually can be carried out by hand as long as the interval of interest is not too large, the development of solution methods could be implemented for some realistic applications even before World War II. Immediately after the war there was a period with more advanced analysis regarding accuracy and stability as we saw in Sect. 4.2. Since ODE solvers by definition deal with only one dimension, it is possible to go into a lot of detail about their behavior, and refined techniques can be introduced without difficulties in their implementation. Here we shall describe some key steps in later developments.

One type of generalization is obtained by increasing the order of accuracy for various known methods. For methods of the backward differentiation type, the generalization is easily derived. By defining the first order difference operator

$$\nabla u_j = u_j - u_{j-1},$$

the BDF of order p for $du/dt = f(t, u)$ is

$$(c_1 \nabla + c_2 \nabla^2 + \ldots + c_p \nabla^p) u_{n+1} = hf(t_{n+1}, u_{n+1}),$$

where h is the stepsize and $c_p \neq 0$. These methods are used for $p \leq 6$; for $p \geq 7$ they are unstable. The methods are A-stable for $p \leq 2$, i.e., for the model equation $du/dt = \lambda u$, the stability domain in the complex plane for $\mu = \lambda h$ is the whole left halfplane $Re\ \mu \leq 0$. Due to the Dahlquist barrier (see Sect. 4.2.2) a linear multi-step method cannot be A-stable for $p \geq 3$, and in these cases the stability domain does not include a certain area to the left of the imaginary axis. This "unstable" area increases with increasing order of accuracy p. BDF methods are much used by modern ODE solvers because they have a damping property which helps when treating non-smooth solutions, but unlike many other types of methods, accuracy is usually limited to the second order variant because of A-stability.

Runge–Kutta methods (from now on called R-K methods) are also popular for modern ODE solvers. The original method used by Runge and Kutta is the fourth order variant, higher order ones were studied in 1925 by Nyström, see [124]. A more general class with the same structure was developed by the New Zealand mathematician John Butcher, whose name is strongly connected with these methods.

These explicit methods have the general form

$$k_1 = f(t_n, u_n),$$
$$k_2 = f(t_n + c_2 h, u_n + h a_{21} k_1),$$
$$\vdots$$
$$k_s = f\big(t_n + c_s h, u_n + h(a_{s1} + a_{s2} k_2 + \ldots a_{s,s-1})\big),$$

$$u_{n+1} = u_n + h \sum_{j=1}^{s} b_j k_j,$$

where the constants a_j, b_j, c_j are specified. Each version is called an s-stage method and can be characterized by the *Butcher tableau*

$$
\begin{array}{c|ccccc}
0 & & & & & \\
c_2 & a_{21} & & & & \\
c_3 & a_{31} & a_{32} & & & \\
\vdots & \vdots & & & & \\
c_s & a_{s1} & a_{s2} & \ldots & a_{s,s-1} & \\
\hline
 & b_1 & b_2 & \ldots & b_{s-1} & b_s
\end{array}
$$

Consistency requires

$$\sum_{j=1}^{i-1} a_{ij} = c_i, \qquad i = 2, 3, \ldots, s,$$

but of course one is interested in conditions for a higher order of accuracy. Here we shall only consider the basic question concerning how many stages are required for obtaining a certain order of accuracy. This is not a trivial task, but Butcher started the investigation in 1963, see [16], and followed up in 1965, see [17]. There are many well known open problems in pure mathematics, but in numerical analysis there are probably fewer. However, here we have one: Given the order of accuracy p, what is the minimum number of stages s to obtain this order? The following table is known today:

p	1	2	3	4	5	6	7	8
$\min(s)$	1	2	3	4	6	7	8	11

The stability domain for explicit R-K methods is necessarily bounded and usually decreasing with a higher order of accuracy. To get better stability properties, implicit versions of R-K methods can be constructed. They have the general form

$$k_i = f(t_n + c_i h, u_n + h \sum_{j=1}^{s} a_{ij} k_j), \qquad i = 1, 2, \ldots, s,$$

$$u_{n+1} = u_n + h \sum_{i=1}^{s} a_{ij} k_j,$$

and require solving nonlinear equations at every stage. We shall not discuss these methods further but note the important fact that unlike linear multistep methods, these methods can be A-stable for any order of accuracy.

ODE solutions may of course have very different degree of smoothness in different parts of the interval of interest. For example, we have seen in Sect. 4.2.2 how a stiff problem may have a very steep initial layer, while the solution later in time is very smooth. This makes it natural to use a different stepsize in different parts of the time-interval. The problem is that in general one doesn't know the behavior of the solution a priori, but an estimate of the truncation error during the solution process might give an indication of the properties of the solution. Bill Gear early on constructed methods that change both the stepsize h and the order of accuracy for the method, and these results are collected in his book [66] published in 1971. Modern general systems containing ODE solvers usually have an automatic choice of stepsize and order of accuracy. In most cases there are certain limits, so that too frequent changes can be avoided.

Bill Gear   (Courtesy Bill
Gear)

In Sect. 3.7.4 we described the principles for constructing accurate methods when a Taylor expansion of the error exists, in particular for ODE solvers. This idea, called the Richardson extrapolation after its inventor, was picked up and generalized in 1963 by William Gragg (1936–2016) (Ph.D. student under the Swiss mathematician Peter Henrici (1923–1987)) in his Ph.D. thesis and later in [78]. His work mainly concentrates on deriving series expansions of the error, and was partly based on the article [15] by Roland Bulirsch and Josef Stoer.

Peter Henrici   (MacTutor
History of Mathematics
archive; http://www-history.
mcs.st-andrews.ac.uk)

So far we have only discussed first order ODE systems. Higher order differential equations including derivatives du^r/dt^r with $r \geq 2$ can always be rewritten as a first order system by introducing new variables

$$v^{(r)} = du^r/dt^r, \qquad r = 1, 2, \ldots,$$

and since a lot is known about such initial value problems, this is often considered as the standard procedure. However, in the article [65] published in 1967, Bill Gear developed methods applied directly to the original high order form, and showed that this is a more efficient technique.

Around 1980 many researchers in nearby areas thought that the theory was fairly complete. In response to this, Gear wrote the article [67] with the title *Numerical solution of ordinary differential equations; Is there anything left to do?* Not surprisingly his answer was that there is. And now in 2018, the same positive answer would certainly be given to the same question. One particular problem is associated with PDEs. A time-dependent PDE is often discretized in space first, resulting in a large system of ODEs. The next step is to apply a finite difference method in time; this procedure is called the *method of lines*. By first analyzing the ODE system one could get an advantage in the analysis by using known results for ODE solvers. However, a new difficulty arises. The ODE system becomes larger without bound with decreasing stepsize in space, but most of the existing analysis has been done under the assumption that the size N of the system is fixed. Here stability and convergence results must be uniform and independent of N.

However, the situation is quite satisfactory when it comes to most problems in science and engineering. Many software packages exist with ODE solvers that are very general and powerful. Not only can the time-step be automatically varied when time proceeds, but also the method itself and its order of accuracy can be varied.

There are several books on the numerical solution of ODEs, and the most well known are written by the most well known names in the area. The book [88] by Henrici came out in 1961 and it covers most of the existing material up to that point in time. The next major one [66] written by Gear came out in 1971 (mentioned above), and it contains many new results that had not been published before.

In 1996 the book [85] by Hairer and Wanner was published with particular emphasis on stiff systems and differential algebraic systems. John Butcher, famous for all his work on Runge–Kutta methods, published the book [18] *Numerical Methods for Ordinary Differential Equations* in 2003, and followed up with a second edition in 2008, and a third edition in 2016.

John Butcher   (Courtesy
John Butcher)

5.4   Initial Boundary Value Problems for PDE and Difference Methods

So far we have discussed pure initial value problems for PDEs, i.e., there is an initial condition in time, and in space, the problem is either defined over the whole domain or alternatively, the solutions are periodic. However, in practice there are most often boundaries somewhere in space, and the Fourier transform cannot be applied. In this section we shall present the key development for constructing stability conditions.

5.4.1   The Godunov–Ryabenkii Condition

In the 1950s the Moscow University group headed by Israil Moyseevich Gelfand began taking an interest in the stability theory for PDE problems with the young student Sergei Godunov (mentioned in Sect. 4.3.5) at the front. The fundamental question was how to find easily applicable conditions for stability, i.e., conditions such that for one-step methods,

$$u(t_{n+1}) = Q(h)u(t_n),$$

$$\|Q(h)^n\| \leq \text{const.},$$

for all h and n. We switch here to h as notation for the stepsize in space, and it is assumed that there is a relation $\Delta t = rh^p$ for some p. The difficulty is that the size of the matrix $Q(h)$ corresponding to the difference operator becomes arbitrarily large since stability must hold for arbitrarily small h. If the eigenvalues of Q for a fixed h are inside the unit circle, the solution tends to zero for $n \to \infty$, but it does not guarantee stability. Figure 5.1 shows what might happen when the grid size decreases.

Consider the simple model problem

$$u_t + u_x = 0, \qquad 0 \leq x \leq 1, \ 0 \leq t,$$

$$u(0, t) = 0,$$

$$u(x, 0) = f(x),$$

and the approximation

$$u_j(t_{n+1}) = u_j(t_n) - r\big(u_j(t_n) - u_{j-1}(t_n)\big), \qquad r = \Delta t/h,$$

$$u_0(t_{n+1}) = 0,$$

$$u_j(0) = f_j.$$

Fig. 5.1 Unstable solution
for decreasing grid size

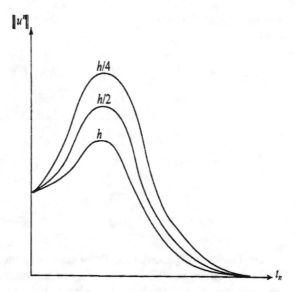

The matrix Q is

$$
Q = \begin{bmatrix}
0 & & & & & \\
r & 1-r & & & & \\
& r & 1-r & & & \\
& & & \ddots & \ddots & \\
& & & & \ddots & \ddots \\
& & & & r & 1-r
\end{bmatrix},
$$

with the two eigenvalues $\lambda(Q) = 0,\ 1-r$, and we have

$$
|\lambda(Q)| \le 1 \quad \Leftrightarrow \quad 0 \le r \le 2.
$$

However, a simple application of the CFL condition discussed in Sect. 3.8.5 shows that this eigenvalue condition cannot be the stability condition. For $r > 1$ the true domain of dependence is not included in the numerical domain of dependence, and therefore it cannot be stable.

This example was constructed by Godunov in 1954. Later he was joined by G.V. Ryabenkii, and they started working on a more general concept that was more suitable for analyzing a *family of operators* rather than a single matrix. This family consists of operators Q_h that depend on the stepsize h. In the paper [71] they defined the spectrum in the following way:

A point λ is a spectral point of the family $\{Q_h\}$ if for any $\varepsilon > 0$ and $h_0 > 0$, there is a number $h,\ h < h_0$ such that the inequality $\|Q_h u - \lambda u\| \le \varepsilon \|u\|$ has a solution u.

Fig. 5.2 Spectrum of Q for
$r = 3/2$

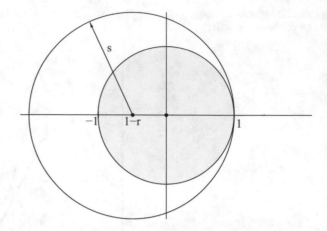

The *Godunov–Ryabenkii condition* states that the whole spectrum lies inside or on the unit circle, and this is a necessary condition for stability. It can be shown that the spectrum for our simple model example is $\lambda = 1 - r + s$ for all complex s with $|s| \leq r$. Figure 5.2 shows the spectrum for the case $r = 3/2$, which violates the Godunov–Ryabenkii condition $0 \leq r \leq 1$. The constant r must be chosen so that the spectrum is contained in the unit disc.

5.4.2   Kreiss' Stability Theory

At the time when the Godunov–Ryabenkii article came out, there was essentially only one way to find sufficient conditions for stability of initial boundary value problems, and that was the energy method. It means that one constructs a scalar product and a norm

$$(u, v)_h = \sum_{j=0}^{N} \alpha_j u_j v_j h, \qquad \|u\|_h^2 = (u, u)_h$$

such that the solution of the difference scheme satisfies

$$\|u(t_{n+1})\|_h \leq \|u(t_n)\|_h, \qquad n = 0, 1, \ldots.$$

But the question is: if one cannot find the coefficients α_j so that this condition is satisfied, is the scheme stable or not? In other words, there is a need for necessary and sufficient stability conditions that can be tested and found to be satisfied or not satisfied. Just as for initial value problems and the von Neumann condition, Kreiss set out to find such a condition. He started out with dissipative approximations, and published his new technique in the 1968 article [100]. Consider the quarter space problem in the domain $x \geq 0,\ \ t \geq 0$,

$$u(t_{n+1}) = Qu(t_n),$$
$$Bu = g, \quad x = 0,$$
$$u(0) = f, \quad t = 0,$$

where B is some operator representing the boundary conditions. Each boundary is treated by itself, so instead of prescribing a boundary condition to the right, the solution is required to be in $L_2(0, \infty)$, which on the discrete side means that $u \in l_2$, i.e.,

$$\sum_{j=0}^{\infty} |u_j(t_n)|^2 h < \infty$$

for all t_n. Furthermore, it is always assumed that the pure initial value problem is satisfied.

There is an eigenvalue problem

$$Q\phi_j = z\phi_j, \qquad \phi \in l_2,$$
$$B\phi = 0, \qquad x = 0$$

associated with the problem. The first equation is an ordinary difference equation in space, and we have seen in Sect. 4.2 that the general solution is determined in terms of the roots κ_ν of the characteristic equation as

$$\phi_j = \sum_\nu \sigma_\nu \kappa_\nu^j(z),$$

where the constants σ_ν arc to be determined. We are interested in the case $|z| > 1$, and it can be shown that the roots of the characteristic equation are separated into two sets, one of them inside the unit circle and the other set outside of the unit circle. Because of the condition $\phi \in l_2$, we keep only the first set, and the coefficients $\sigma = [\sigma_1, \sigma_2, \ldots, \sigma_r]$ are determined by the boundary conditions. This leads to a system of equations

$$E(z)\sigma = 0,$$

where $E(z)$ is an $r \times r$ matrix, and there is an eigenvalue z_0 if the condition

$$\mathrm{Det}\big(E(z_0)\big) = 0 \tag{5.1}$$

is satisfied.

The crucial question is what happens if z approaches the unit circle. For dissipative approximations, it turns out that at any point except for $z = 1$, all the roots κ_ν stay inside the unit circle, but at $z = 1$, one or more of the roots fall on the unit circle. But then the norm of the corresponding eigenvector ϕ is no longer finite, and therefore we call $z = 1$ a *generalized eigenvalue* if

$$\mathrm{Det}\big(E(1)\big) = 0.$$

Kreiss's main result was that a dissipative difference scheme is stable if there is no eigenvalue with $|z| > 1$ or generalized eigenvalue $z = 1$.

For most real life problems there is another boundary, for example at $x = 1$. To ensure stability we require that also the left quarter space problem defined in the domain $-\infty < x \leq 1$ be stable.

Kreiss worked also on the initial boundary value problem for hyperbolic PDE systems without any discretization. There was no general stability theory for this problem, but in [101] he presented a complete theory based on a Laplace transformation in time. The same technique could be modified for difference equations as well, and in [82] it was applied to general multistep schemes. The definition of stability is somewhat different there than in the traditional one. The norm of the solution is defined as the l_2-norm taken over both space and time, and the estimate is in terms of the boundary data. The technique for checking the stability condition is the same as above, but now the whole unit circle must be checked for generalized eigenvalues. The condition for stability can be formulated as the *determinant condition*

$$\mathrm{Det}\big(E(z)\big) \neq 0, \qquad |z| \geq 1. \tag{5.2}$$

It can be shown that the Godunov–Ryabenkii condition is the same condition, but relaxed to the reduced domain $|z| > 1$.

The theory goes sometimes under the name GKS-theory after the initials of the authors, but clearly, Kreiss was the father of the principles and of the main work. Checking the stability condition is also called normal mode analysis, since one can formally use the ansatz

$$u_j(t_n) = \kappa^j z^n \tag{5.3}$$

for complex scalars κ and z.

Instead of presenting the rather complicated theory, we give an example that illustrates how the stability condition is applied. Consider the quarter space problem

$$\frac{\partial u}{\partial t} = \frac{\partial u}{\partial x}, \qquad x \geq 0, \ t \geq 0, \tag{5.4}$$

and the leap-frog scheme

$$v_j(t_{n+1}) - v_j(t_{n-1}) = \lambda\big(v_{j+1}(t_n) - v_{j-1}(t_n)\big),$$

where $\lambda = \Delta t / h$. Since the characteristics of (5.4) are propagating from right to left, this PDE requires no boundary condition at $x = 0$. However the leap-frog scheme does, and we choose a simple second order extrapolation

$$v_0(t_{n+1}) = 2v_j(t_{n+1}) - v_2(t_{n+1}).$$ (5.5)

By using the ansatz (5.3), the characteristic equation is obtained as

$$\kappa^2 - \frac{z^2 - 1}{\lambda z} \kappa - 1 = 0,$$

and for $|z| > 1$, the roots satisfy $|\kappa_1| < 1$, $|\kappa_2| > 1$. The question is now how κ_1 behaves when z approaches the unit circle $e^{i\theta}$. It is easily shown that

$$\kappa_1 = \frac{\lambda \pm (\lambda^2 - \sin^2 \theta)^{1/2} + i \sin \theta}{\lambda \pm (\lambda^2 - \sin^2 \theta)^{1/2} - i \sin \theta},$$

where the sign is to be chosen opposite to that of $\cos \theta$. For the dissipative case, the only critical point is $\theta = 0$, and here we have $\kappa_1 = -1$. The matrix E in (5.1) is a scalar $(\kappa_1 - 1)^2$, which shows that κ_1 is not a generalized eigenvalue. However, here we have also another critical point $\theta = \pi$, i.e., $z = -1$, where $\kappa_1 = 1$. Accordingly, $E(-1) = (\kappa_1 - 1)^2 = 0$, and we have a generalized eigenvalue $z = -1$. The conclusion is that the leap-frog approximation is unstable with the boundary condition (5.5). It is now easy to construct another boundary condition

$$v_0(t_{n+1}) = 2v_1(t_n) - v_2(t_{n-1})$$

providing stability. We have $E(z) = (z - \kappa)^2$, and since it never happens that $z = \kappa_1$ for $|z| \geq 1$, we have $E(z) \neq 0$, which implies that there is no eigenvalue or generalized eigenvalue.

The stability theory reduces the analysis to solving the characteristic equation, which is a polynomial in κ in terms of z, and then investigating whether or not the matrix $E(z)$ is singular. The procedure is a little bit more complicated than checking the von Neumann condition, but still doable. Furthermore, the theory guarantees that the determinant condition (5.2) is *equivalent* to stability, not only necessary or only sufficient.

5.5 New Types of Functions

For centuries there were essentially two types of functions used when numerical methods were constructed: polynomials and trigonometric functions. But new types of functions have been constructed during the last decades, often motivated by a need for more effective numerical methods. In this section we shall discuss some of the more important ones.

5.5.1   Piecewise Polynomials and Splines

In Sect. 3.2 we discussed the problem of using polynomials for interpolation. If there are many data points, polynomials usually become very bad approximations in-between these points. A remedy for this is to define polynomials in each subinterval, and then connect them in some way. As an example, we consider the case with uniformly distributed points in the interval $[a, b]$,

$$x_j = a + jh, \qquad j = 0, 1, \ldots, N, \quad Nh = b - a.$$

A piecewise polynomial $P_n(x)$ of degree n with nodes x_j is a function that is a polynomial of degree n on each subinterval $[x_j, x_{j+1}]$. The regularity at the nodes is limited and can be chosen in different ways; continuity is in most cases a minimal requirement.

Piecewise linear functions is an obvious and easy choice for interpolation. On each subinterval, a linear function is determined by the value of the given function $f(x)$ at the two endpoints, and it automatically becomes continuous at the nodes. Actually, Archimedes had already used piecewise linear functions when approximating the circle. For higher order polynomials we need extra points inside the interval in order to get a unique function. Let us take piecewise cubics as an example. Two inner points $x_{j+1/3}$ and $x_{j+2/3}$ are needed, and assuming that these function values are given, the cubic is determined, and the piecewise polynomial is continuous everywhere. There is a difference here between the inner extra points and the original nodes, since the function is infinitely differentiable at the inner ones.

If the derivatives $f'(x_j)$ at the nodes are given as well, no extra inner points are necessary. The four coefficients are determined by f and f' at the endpoints. This is Hermite interpolation on each subinterval, as mentioned in Sect. 3.2. However, there is a third way to obtain a piecewise cubic, which does not require any extra knowledge about the function at the nodes. The extra two equations for determinating the coefficients can be obtained by requiring that the first and second derivatives match the same quantities of the local cubic to the left. This gives a function $P_n(x)$ with continuous derivatives up-to-order two everywhere in $[a, b]$. It is called a *spline* and was first introduced in 1946 by Isaac Schoenberg (1903–1990), see[149].

It can be shown that this is the interpolation function that satisfies the minimization problem

$$\min_{g(x)} \int_a^b |g''(x)|^2 \, dx.$$

Actually, there is a physical interpretation of this problem and its solution. In the old days, engineers wanted to make a drawing from a set of given points on a contour. A flexible ruler was forced to pass through the given points, and the resulting shape of the ruler was copied. The result was the cubic spline.

The spline concept can be further generalized to any degree of interpolating polynomials $P_n(x)$ with the extra condition that

$$|\frac{d^{n-1} P_n}{dx^{n-1}}(x)|$$

be uniformly bounded on the whole interval $[a, b]$.

After the Schoenberg initial article on splines, a lot of work and development was done by others, and many researchers were involved. Carl de Boor is among the leaders, and his important article [37], published in 1972, about the technique for evaluating splines was followed by his book [38] in 1978, which is a complete presentation of all the essentials concerning this topic. In practice one often needs interpolation in several dimensions. As long as we are dealing with uniform grids, the techniques described above for 1-D problems can be quite easily generalized to multi-D as long as we use Cartesian product spaces, i.e., rectangular grids in 2-D and correspondingly for higher dimensions. However, for an arbitrary set of points in several dimensions, the problem becomes more difficult.

Constructing grids and interpolating piecewise polynomials actually came out as a biproduct of the fast development of finite element methods that took place beginning in the 1950s. We shall discuss this development in Sect. 5.6, but bring out the basics as it applies to interpolation here.

The natural generalization of piecewise linear interpolation from 1-D to 2-D is not by going to structured rectangular grids. The reason is that a linear function

$$P_1(x, y) = a_{10}x + a_{01}y + a_{00}$$

is uniquely determined by three points in the (x, y)-plane, and a rectangle has 4 corners. In the latter case we need the extra term $a_{11}xy$ to give a bilinear function. Therefore triangles are the right choice for linear functions. This is convenient also from another point of view. For irregular domains, triangles are much better when it comes to approximating the boundary by the triangle sides. In 1943 Courant had already used triangular partition of the domain in his first trial with finite elements, as discussed in Sect. 3.8.2 (Figs. 5.3 and 5.4).

When going to second degree polynomials, triangles are again the natural choice. There are now 6 coefficients to be determined and consequently, if a node is added in the middle of each edge, we have the right number. For cubic polynomials, there are 10 coefficients. By adding two nodes at each edge, we get 9 nodes altogether, and we need one more. This one is chosen as the center of gravity of the triangle, see Fig. 5.5.

In one dimension the interpolating polynomials are continuous across the nodes, since we require that the polynomials on each side equal the given function values,

Carl de Boor   (Courtesy Carl de Boor)

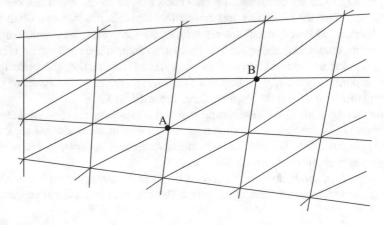

Fig. 5.3 Triangular grid

and as a consequence, they equal each other at the common point. In 2-D the continuity doesn't follow immediately. Two different polynomials have the same value at the common nodes, but do they agree along the whole edge? Indeed they do.

Consider the two triangles in Fig. 5.6 with polynomials $P_1(x, y)$ and $P_2(x, y)$ respectively.

The triangle side AB is defined by an equation $y = \alpha + \beta x$ for some constants α and β, and when replaced into the second degree polynomials on each side of the triangle edge, we obtain two quadratic polynomials in the single variable x. These are uniquely determined by the function values at the three points A,B,C, showing that they are identical, and consequently there is continuity at every point.

The same conclusion is easily made for cubic polynomials on each side if there are two inner nodes on each triangle side.

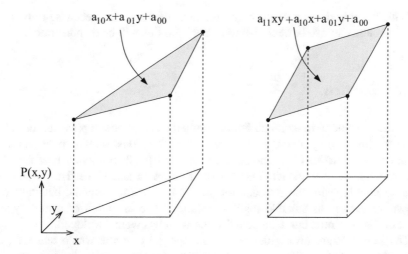

Fig. 5.4   Linear and bilinear functions

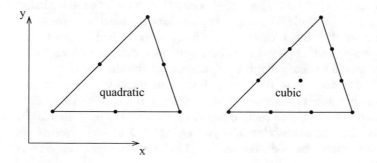

Fig. 5.5   Nodes for quadratic and cubic polynomials

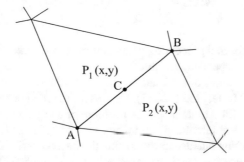

Fig. 5.6   Two triangles with different polynomials

With the introduction of the finite element method, the piecewise polynomial function class came to be central, and in Sect. 5.6 this will be demonstrated.

5.5.2   Wavelets

A central problem in computational mathematics is the representation of an unknown function by a combination of known functions, such as polynomials or trigonometric functions. The choice is governed by the flexibility of these functions and the amount of work required for their computation. Piecewise polynomials became popular with the introduction of finite element methods, but they were known very early as suitable for interpolation. The invention of a new type of functions is very rare, but it happened when wavelets were developed.

The Fourier transform is defined as an integral over the whole real line, and is suitable for finding the distribution of different frequencies. It is no good for the analysis of functions that are characterized by shorter pulses. A remedy for this limitation is some sort of localized Fourier transform as suggested in 1946 by Dennis Gabor (1900–1979), see [59]. The transform is taken in a window which is gliding over the whole line, in that way catching the local behavior of the function. A remaining difficulty is caused by the fact that the typical behavior of the function is not known, and the proper size of the window is difficult to find. A more efficient method would be obtained if this limitation could be eliminated.

The research efforts along these lines were mainly carried out in the signal processing community. The result was a new set of functions, later called *wavelets*. The origin can be traced back to the Ph.D. thesis 1910 by Alfred Haar, see [83]. He was a Hungarian mathematician getting his Ph.D. at the University of Göttingen under the supervision of the famous David Hilbert. The new invention was using the basis function

$$\Psi(x) = \begin{cases} 1, & 0 \leq x < 1/2, \\ -1, & 1/2 \leq x < 1, \\ 0, & \text{else} \end{cases}$$

shown in Fig. 5.7. (Signal processing deals naturally with time, but here we use the notation x for the independent variable.)

Various generalizations of the Haar concept were done, but the real leap forward was made by Jean Morlet (1931–2007). He was a French geophysicist who worked for the oil company Elf Aquitaine on the problem of analyzing echoes coming back from sound signals sent into the ground when searching for oil reservoirs. He found that by not only shifting but also transforming the basis functions such that different scales were introduced, a much better result was obtained. The multiresolution concept was invented, and it led to the construction of more general wavelets.

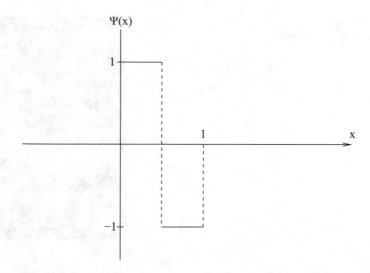

Fig. 5.7 The Haar function

If $\Psi(t)$ is the original basis wavelet, for example the Haar wavelet above, the transformed wavelets are characterized by the scaling parameter a and the translation parameter b in the form

$$\Psi_{a,b}(x) = \Psi\left(\frac{x-b}{a}\right).$$

The theoretical analysis containing the wavelet transformation and its inverse was worked out by Jean Morlet together with Alex Grossman, and the result was presented in 1984, see [80]. This started many new activities in the mathematical community. Wavelet analysis became a standard tool in various applications, and new basis functions were constructed.

To be of any practical use for computation, the wavelet transform has to be discretized, just like the Fourier transform. A particularly important work in this respect was done by Ingrid Daubechies, who got her Ph.D. in quantum mechanics at the Vrije University of Brussels in 1980. She worked with Grossman, but moved to the United States where she was employed by AT&T Bell Laboratories. She had access to the early work of Stephane Mallat and Yves Meyer in multiresolution analysis. In 1988, Daubechies came up with a complete theory for discrete wavelets connected to multiresolution analysis, see [35]. Briefly it goes as follows.

Ingrid Daubechies
(Courtesy Ingrid Daubechies)

The common choice for scaling and translation parameters is $a = 2$ and $b = 1$, and this is what we choose here. Each wavelet is defined by a *scaling function* (or father function) ϕ and two filters with coefficients h_k and g_k. The scaling function satisfies the equation

$$\phi(x) = \sqrt{2} \sum_{k=0}^{\infty} h_k \phi(2x - k),$$

and is normalized such that

$$\int_{-\infty}^{\infty} \phi(x)\, dx = 1.$$

The wavelet Ψ is then

$$\Psi(x) = \sqrt{2} \sum_{k=0}^{\infty} g_k \phi(2x - k).$$

For the Haar wavelet, the scaling function is the unit box function

$$\phi(x) = \begin{cases} 1, & 0 \leq x < 1, \\ 0, & \text{else}, \end{cases}$$

and the nonzero filter coefficients are

$$h_0 = h_1 = 1/\sqrt{2},$$

$$g_0 = 1/\sqrt{2}, \quad g_1 = -1/\sqrt{2}.$$

The basis functions are obtained by scaling and translation as

$$\phi_{jk} = 2^{-j/2}\phi(2^{-j}x - k),$$
$$\Psi_{jk} = 2^{-j/2}\Psi(2^{-j}x - k).$$

The filters are now chosen so that the basis functions are orthonormal. A given function can then be represented in the form

$$f(x) = \sum_j \sum_k c_{jk}\Psi_{jk}(x),$$

where

$$c_{jk} = \int_{-\infty}^{\infty} \Psi_{jk}(x)f(x)\,dx. \tag{5.6}$$

It is trivial to show that the Haar function provides an orthonormal basis. Daubechies generalizes this concept of orthonormal basis functions to a higher approximation order, and furthermore, she constructs them such that each one has compact support. If the first p moments vanish, i.e.,

$$\int_{-\infty}^{\infty} x^j\Psi_{jk}(x)\,dx, \qquad j = 0, 1, \ldots, p - 1,$$

then polynomials of degree $p - 1$ are exactly represented, and the approximation order is p. Furthermore, the filters are chosen with finite length, such that the second condition is satisfied as well. Accuracy of order p is obtained with filter length p, i.e., there are p filter coefficients of each kind.

We have seen above that the Daubechies wavelet with $p = 2$ is the Haar function with the simple structure of ϕ and Ψ. But for $p = 4$ we already have a very different situation. Indeed, the filter coefficients can be found analytically as

$$h_0 = \frac{1 + \sqrt{3}}{4\sqrt{2}}, \quad h_1 = \frac{3 + \sqrt{3}}{4\sqrt{2}}, \quad h_2 = \frac{3 - \sqrt{3}}{4\sqrt{2}}, \quad h_3 = \frac{1 - \sqrt{3}}{4\sqrt{2}},$$

$$g_0 = h_3, \quad g_1 = -h_2, \quad g_2 = h_1, \quad g_3 = -h_0,$$

but the scaling and wavelet functions cannot be defined by any explicit expression. However, they can be generated to any order of accuracy by the Daubechies machinery. The shape of these functions may have come as a surprise for Daubechies as well who comments on them in her article when writing "*There are several striking features...* ". They are quite irregular in shape and there is no symmetry, see Fig. 5.8.

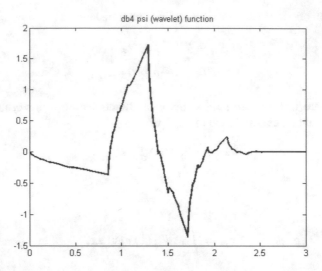

Fig. 5.8 Daubechies wavelet of order 4 (generated by Jonny Latuny)

Her paper [35] is very well written with many examples, but for a more general description of the whole topic, Daubechies' book published in 1992 is recommended, see [36].

In addition to the good approximation properties of these functions, they are convenient also from the point of view that the wavelet transform can be computed quickly. There is no need to approximate the integrals (5.6) that theoretically define the coefficients of the expansion for any orthonormal basis. Instead they can be computed fast by an algorithm that is closely connected to the derivation of the filter coefficients. In the article [9] fast algorithms based on wavelet transforms are developed for applying them to a number of applications that after discretization can be expressed as matrix-vector multiplications.

Wavelets have their origin in the signal analysis community. Image analysis is another area where the impact has been significant, since wavelets have the ability to represent details of images on widely different scales. This new type of approximating functions also naturally led to attempts to use them for numerical solutions of differential equations, in particular partial differential equations. These problems often have solutions that vary on many different scales, being smooth in some areas but having severe irregularities like shocks in other areas. Galerkin-, spectral- or pseudo-spectral methods can be used with wavelet bases. In the latter case, wavelets have a potential of being much more effective than Fourier expansions (as described in Sect. 5.8) for large classes of problems, and in the 1990s the use of wavelets for PDEs gained momentum. However, even if these methods have some use for a few applications, they have not become very popular. One reason is that computing derivatives is not so simple. For example, in the spectral method setting, this computation is trivial for Fourier expansions, where differentiation is a simple multiplication of the coefficients by a scalar. For wavelets, differentiation is a much more complicated operation. Another drawback, which is shared with Fourier methods, is the difficulty of implementing boundary conditions for nonuniform domains.

5.5.3 Radial Basis Functions

For problems in d space dimensions, we have so far always worked with functions of d different coordinates that are independent of each other. For large d, the common technique for using a uniform distribution of grid-points in each direction is quite restrictive and leads to heavy computations. Nonuniform grids can be introduced as in finite element methods, but even more flexibility is obtained if the grid can be abandoned altogether, and place the nodes in a way that is most natural for the particular application in question. In 1971, Rolland Hardy came up with such a type of functions which was later generalized and called radial basis functions. Such a function is a function $\phi(r)$ of only one variable, namely the Euclidean distance r from the origin.

Let us now see how these functions are used for approximating a function with given values at a finite number of points. We use the notation $\mathbf{x}$ for a point in the d-dimensional space R^d with coordinates

$$\mathbf{x} = \begin{bmatrix} x^{(1)} \\ x^{(2)} \\ \vdots \\ x^{(d)} \end{bmatrix}.$$

Assuming that a function $f(\mathbf{x})$ is given at N points $\mathbf{x}_1, \mathbf{x}_2, \ldots, \mathbf{x}_N$, we look for an approximation

$$s(\mathbf{x}) = \sum_{j=1}^{N} \lambda_j \phi(\|\mathbf{x} - \mathbf{x}_j\|),$$

where the coefficients λ_j are to be determined. The basis functions are defined via one single function ϕ, but are distinguished by having N different center points $\mathbf{x}_j$. For straightforward interpolation, we add the obvious conditions

$$s(\mathbf{x}_j) = f(\mathbf{x}_j), \qquad j = 1, 2, \ldots, N.$$

It is not certain that the interpolation problem has a unique solution; furthermore the resulting algebraic system may be difficult to solve efficiently. However, during the last two decades, this problem has been a topic for quite intensive investigations, and several effective algorithms are available.

The original function ϕ suggested by Hardy was $\phi(r) = \sqrt{1 + r^2}$ and is called *multiquadratic*, but many other functions have been introduced. It is quite natural to construct them such that they tend to zero for large r. Furthermore it was soon realized that a scaling parameter ε should be introduced. Some commonly used functions $\phi(r, \varepsilon)$ are

$$\frac{1}{\sqrt{1 + (\varepsilon r)^2}}, \qquad \frac{1}{1 + (\varepsilon r)^2}, \qquad e^{-(\varepsilon r)^2}.$$

Radial basis functions can also be used for solving partial differential equations, and all the numerical methods based on a set of basis functions can be used. Spectral methods are the most natural, but finite element methods are also possible, both described later in this book. In the latter case, the common piecewise polynomials are replaced by radial basis functions with compact support. Collocation methods is another way, and we shall briefly describe an example which was presented by Elisabeth Larsson and Bengt Fornberg, see [107].

Let Δ denote the Laplace operator in d dimensions defined by

$$\Delta u(\mathbf{x}) = \frac{\partial^2 u}{\partial (x^{(1)})^2} + \frac{\partial^2 u}{\partial (x^{(2)})^2} + \ldots + \frac{\partial^2 u}{\partial (x^{(d)})^2},$$

and consider the Dirichlet problem in a domain Ω with boundary $\partial\Omega$:

$$\Delta u(\mathbf{x}) = F(\mathbf{x}), \qquad \mathbf{x} \in \Omega,$$

$$u(\mathbf{x}) = g(\mathbf{x}), \qquad \mathbf{x} \in \partial\Omega.$$

The solution is represented at a set $\mathbf{x}_{\text{Int}}$ with N points in Ω and another set $\mathbf{x}_B$ with M points on the boundary $\partial\Omega$:

$$\mathbf{x}_1, \mathbf{x}_2, \ldots, \mathbf{x}_N, \qquad \mathbf{x}_{Int} \in \Omega,$$

$$\mathbf{x}_{N+1}, \mathbf{x}_{N+2}, \ldots, \mathbf{x}_{N+M}, \qquad \mathbf{x}_B \in \partial\Omega.$$

The straightforward collocation method introduced by E.J. Kansa, see [95], is

$$\sum_{j=1}^{N+M} \lambda_j \Delta\phi(\|\mathbf{x}_k - \mathbf{x}_j\|, \varepsilon) = F(\mathbf{x}_k), \qquad k = 1, 2, \ldots, N,$$

$$\sum_{j=1}^{N+M} \lambda_j \Delta\phi(\|\mathbf{x}_k - \mathbf{x}_j\|, \varepsilon) = g(\mathbf{x}_k), \qquad k = N+1, N+2, \ldots, N+M,$$

$$(5.7)$$

giving the solution

$$s(\mathbf{x}, \varepsilon) = \sum_{j=1}^{N+M} \lambda_j \phi(\|\mathbf{x} - \mathbf{x}_j\|, \varepsilon).$$

It has been shown that the system (5.7) may be singular in exceptional cases, but is still often used. There are other variants of the collocation methods, some of them guaranteeing nonsingularity.

Radial basis functions have their strength when solving problems with many coordinate directions. One such case is often occurring in financial mathematics (see Sect. 5.14.1), where the methods have been refined considerably. An example with application to a time-dependent problem is described in the article [108].

5.6 Finite Element Methods

Partial differential equations may be labelled as the largest class of mathematical models for problems in science and technology. Except for very simple model problems, it is impossible to solve them in terms of known explicit functions that can be evaluated easily. We have described finite difference methods that are based on a direct approximation of derivatives by divided differences. In the 1950s a completely new class of methods showed up, namely finite element methods (FEM). Today they dominate the PDE arena.

5.6.1 Structural Mechanics

In Sect. 3.8.2 we described the first occurrence of a finite element method (but not called so) in an article by Courant, in which it was demonstrated for a torsion problem. However, it was not until the beginning of the 1950s that meaningful FEM-computations took off. Also then, problems in structural mechanics were the main focus, primarily those originating from the aircraft industry. This should not come as a surprise. When constructing aircraft, both military and commercial, it is extremely important to keep the weight of the structure down while keeping the tenacity at a sufficiently high level. The equations of structural mechanics are complicated, and the complicated geometry of aircraft makes it impossible to optimize the structure using engineering rules of thumb. The Boeing company in Seattle started using computational methods very early. Jon Turner was one of the leading engineers there, and around 1950 he began using subdivisions of the wings and applying finite element methods. He almost certainly did this without knowing about Courant's work, which was published in a mathematical journal. Furthermore, he developed FEM for application to more general problems, and it seems safe to claim that he was the inventor of FEM as a general tool for solving problems in structural mechanics. Industrial engineers don't publish very often in academic journals, and the article [155] published by Turner and his coworkers in 1956 seems to be the first publication. One of the coauthors was R.W. Clough at the University of California, Berkeley, who in 1960 for the first time introduced the concept of finite elements, see [22]. In a later article [23] Clough acknowledged Turner by writing *"Also it should be recognized that the principle credit for conceiving the procedure should go to M.J. Turner, who not only led the developmental effort for the two critical years of 1952–1953, but who also provided the inspiration to use assumed strains to define the stiffness of triangular plane stress elements"*.

Thus, the start of FEM as the standard method for PDE problems was driven by engineering problems in structural mechanics. The challenge was to go from structures consisting of connected distinct parts like rods and plates to general continuous structures. The key was to partition such structures into artificial substructures. In each of them, physical functions like stress are represented by linear functions or, more generally, by polynomials. This allows for a much greater flexibility than the alternative of having representation by polynomials defined over the whole domain. The solution is represented by piecewise polynomials instead of polynomials. This is a kind of discretization, but in contrast to finite difference methods, the approximate solutions are represented by functions defined everywhere.

On the other hand, we encounter another difficulty. When connecting polynomials across element boundaries, regularity goes down. With first degree polynomials, the functions can be forced to be continuous across element boundaries, but the derivatives are not. Second derivatives are present in the differential equations, so how are they defined? The remedy is to reformulate the problem so that the regularity of the functions can be relaxed.

As we saw in Sect. 5.5.1, triangles are convenient subdomains in 2-D when it comes to continuity across interfaces and to approximating irregular boundaries. The same advantages are obtained with tetrahedrons in three dimensions.

5.6.2 General Stationary Problems

In the mathematical society it took some time before FEM attracted much interest. In addition to Clough, there was another person from the academic world who had a certain contact with the Boeing group, namely the Greek mathematician John Hadji Argyris. He worked at the Imperial College in London when he published the article [2] in 1954. Argyris then became a main player in the FEM arena with a series of papers in the 1960s and 1970s.

In the 1960s FEM started attracting great interest from many other mathematicians. One reason may have been that in addition to FEM's ability to solve complicated problems in engineering applications, it has a striking simple beauty when formulated in abstract mathematical terms. Let us take Courant's example in Sect. 3.8.2 as an illustration.

In Sect. 3.8.1 we described the Ritz–Galerkin method, and we summarize it here for the special case of finite elements. We look for the solution of an elliptic problem which is reformulated as a minimization problem:

$$\min_{v} I(v), \qquad I(v) = a(v, v) - 2(F, v),$$

where $a(u, v)$ is a positive definite bilinear form with

$$a(v, v) \geq \sigma \|v\|^2, \qquad \sigma > 0$$

for some norm $\| \cdot \|$. The admissible functions v belong to some function space $\mathscr{S}$, and the minimum is given by the function u that satisfies

$$a(u, v) = (F, v) \quad \text{for all } v \in \mathscr{S}. \tag{5.8}$$

The bilinear form $a(u, v)$ is connected to a differential operator L, and for smooth functions u, v satisfying the boundary conditions, we have

$$a(u, v) = (Lu, v).$$

This means that the solution satisfies the differential equation

$$Lu = F,$$

which is called the Euler equation.

In the Courant example we have

$$L = -(\partial^2/\partial x^2 + \partial^2/\partial y^2),$$

$$a(u, v) = \int\int_{\Omega} (u_x v_x + u_y v_y) dx\, dy,$$

$$F = -1,$$

$$\mathscr{S} = \{v / \int\int_{\Omega} (|v_x|^2 + |v_y|^2 + |v|^2) dx dy < \infty, \qquad v = 0 \text{ on the boundary } \Gamma\},$$

i.e., the original problem is

$$u_{xx} + u_{yy} = 1, \qquad u \in \Omega,$$

$$u = 0, \qquad u \in \Gamma.$$

The problem can now be stated in an abstract form as
* *Find the function $u \in \mathscr{S}$ such that (5.8) is satisfied for all $v \in \mathscr{S}$.*

For an approximate solution of the same problem, we define a subspace approximation $\mathscr{S}_h \subset \mathscr{S}$, where the subscript h indicates a discretization parameter (in Sect. 3.8.1 we used the notation $\mathscr{S}_N$ with N denoting the number of basis functions). In the Courant example, we have a partition of Ω into triangles, and then $\mathscr{S}_h$ consists of all functions v_h satisfying the following three conditions:

1. v_h *is linear on each triangle,*
2. v_h *is continuous across triangle edges,*
3. $v_h = 0$ *on* Γ.

The approximate solution u_h is then defined:
* *Find the function* $u_h \in \mathscr{S}_h$ *such that*

$$a(u_h, v_h) = (F, v_h) \text{ for all } v_h \in \mathscr{S}_h.$$

Such an elegant formulation! It is exactly as the original weak formulation, and it only remains to construct the approximation space $\mathscr{S}_h$. This fact is not only beautiful by itself, but also helps to analyze the properties of the numerical solution. It turns out that under certain conditions the error $\|u - u_h\|$ in the solution is of the same order as that of the pure approximation error, i.e., the "distance" between $\mathscr{S}_h$ and $\mathscr{S}$.

The space $\mathscr{S}_h$ is spanned by a finite number of basis functions ϕ_j, $j = 1, 2, \ldots, N$. In the piecewise linear case on triangles, these basis functions are nonzero on a few connecting triangles, and zero in the remaining domain. Figure 5.9 shows a triangular grid with the shaded area representing the part where a certain basis function is nonzero.

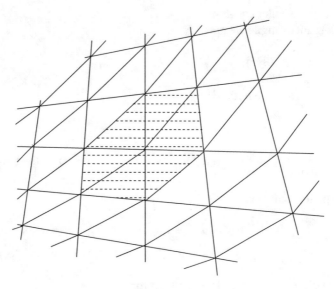

Fig. 5.9 Nonzero support for one basis function ϕ_j

The FEM solution is then defined:
* *Find the function*

$$u_h = \sum_{j=1}^{N} c_j \phi_j \tag{5.9}$$

such that

$$a(u_h, \phi_j) = (F, \phi_j), \qquad j = 1, 2, \ldots, N.$$

This leads to a sparse system of algebraic equations for the unknown coefficients c_j, and some of the methods described earlier can be applied for its solution. Actually, direct methods of the Gauss type were used quite a lot in the beginning, but have later been replaced by iterative methods.

5.6.3 Time-Dependent Problems

After the establishment of FEM as a useful and practical method for problems in structural mechanics in the 1950s, FEM caught the interest of mathematicians and other scientists from various applied areas. The British mathematician and engineer Olgierd Zienkiewicz (1921–2009) was one of them. He had a Polish father and the family moved from England to Poland when he was a boy. He would probably have stayed there, but here we have again an example where the Nazis caused a scientist to leave the European mainland. When Poland was invaded, the family managed to escape and moved to France only to find the German troops attacking this country as well. They escaped once more and reached Great Britain in the summer of 1940, and this is where his scientific career took place, later with a chair at the University of Swansea.

In the 1950s he started working with Turner on engineering problems at Boeing, and he was one of the first to develop the method for more general problems. His article [163] has the general title "Finite elements in the solution of field problems", so here the name "finite elements" had been established.

The 1960s saw the first applications of FEM to initial boundary value problems for time-dependent PDEs. It was natural to first apply them to parabolic problems,

$$u_t = Lu + F,$$

where L is an elliptic differential operator in space. From the beginning, the FEM discretization was applied only to the space part of the equation. We take a heat conduction problem in 2D as an example:

$$u_t = \big(c(x, y)u_x\big)_x + \big(c(x, y)u_y\big)_y + F(x, y), \qquad (x, y) \in \Omega,$$

$$u(x, y, t) = g(x, y, t), \qquad (x, y) \in \Gamma, \qquad (5.10)$$

$$u(x, y, 0) = f(x, y).$$

Here the heat conduction coefficient is positive with $c(x, y) \geq \delta > 0$. In order to simplify the analysis, we choose $F = 0$ and $g = 0$, multiply by a function $v = v(x, y)$, integrate over Ω and use Green's theorem. The result is the weak form

$$(u_t, v) = -(cu_x, v_x) - (cu_y, v_y), \qquad v \in \mathscr{S},$$

and by choosing $v = u$, we get

$$\frac{1}{2}\frac{d}{dt}\|u\|^2 = -\|cu_x\|^2 - \|cu_y\|^2 \leq -\delta(\|u_x\|^2 + \|u_y\|^2).$$

Integrating over time, we get for any t,

$$\|u(t)\|^2 \leq \|f\|^2 - 2\delta \int_0^t \big(\|u_x(\tau)\|^2 + \|u_y(\tau)\|^2\big)d\tau.$$

This shows the well known property of parabolic problems; the norm of the solution is decreasing with time as long as the solution is not constant. But the big effect happens at the next stage. The approximate solution is obtained from

$$\big((u_h)_t, v_h\big) = -\big(c(u_h)_x, (v_h)_x\big) - \big(c(u_h)_y, (v_h)_y\big), \qquad v_h \in \mathscr{S}_h. \qquad (5.11)$$

Just because the approximate solution u_h is defined in exactly the same way as u, and furthermore $\mathscr{S}_h \subset \mathscr{S}$, the same inequality holds:

$$\|u_h(t)\|^2 \leq \|f\|^2 - 2\delta \int_0^t \big(\|(u_h)_x(\tau)\|^2 + \|(u_h)_y(\tau)\|^2\big)d\tau.$$

Not only do we get a stable numerical solution, but this solution also has the same smoothing properties as the true solution. This is in contrast to finite difference methods, in which the stability analysis requires a quite complicated theory.

Equation (5.11) is a system of ordinary differential equations for the coefficients of piecewise polynomials, and it is usually referred to as the method of lines. It is not a complete numerical method, since it requires a numerical ODE solver. A difference method applied in time is an obvious way of completing discretization. However, there is an inherent problem here.

Let **c** denote the vector containing the coefficients c_j in the expansion (5.9). Then the system of ODEs can be written in the form

$$M \frac{d\mathbf{c}}{dt} = K\mathbf{c},$$

where M is a nondiagonal matrix (called the mass matrix). This means that even if an explicit difference method is used for time-discretization, the method is implicit in the sense that it requires the solution of a system of equations for each timestep. As a consequence, one may as well use implicit methods, such as the trapezoidal rule

$$(M - \frac{\Delta t}{2} K)\mathbf{c}(t_{n+1}) = (M + \frac{\Delta t}{2} K)\mathbf{c}(t_n),$$

which has a structure similar to the Crank–Nicolson difference scheme described earlier. So, even if also the FEM version used here is unconditionally stable, the coupling caused by the mass matrix is in general a disadvantage. But there is a way of avoiding this and still retain the finite element structure.

5.6.4 Discontinuous Galerkin Methods

As demonstrated above, the FEM theory requires that approximating functions have a certain regularity across element boundaries. In the simplest 2-D example with triangles and piecewise linear functions with nodes at the corners, continuity is guaranteed along all sides of the triangle. For higher order approximations and higher order derivatives in the differential equation, stricter regularity properties are required. Furthermore, as seen above, one effect of time-dependent problems is that any time-discretization becomes implicit since all unknowns in the discrete system become coupled to each other. The question now is if this coupling can be removed without any serious consequences.

The Galerkin method is based on an approximation subspace that is spanned by globally defined basis functions ϕ_j. Even if they are zero outside of a subdomain consisting of a small number of triangles (in 2-D), they are well defined and at least continuous in the whole computational domain. If instead they are defined locally on a small subdomain such that the solution is discontinuous across the boundaries between the subdomains, there is the possibility of obtaining a more loosely coupled system.

This new approach was introduced by Reed and Hill in a Los Alamos report [137] with the title *"Triangular mesh methods for the neutron transport equation"* published in 1973. This equation is a first order hyperbolic PDE in space without any time-dependence. The authors make a point of not only replacing the usual rectangles with triangles for discretization of such a problem, but they also note that the usual Galerkin formulation leads to a coupling of all variables. This is a real disadvantage, since a hyperbolic PDE has a structure with propagation along characteristics allowing for explicit time-stepping. Their solution is to define polynomials locally on each triangle without any continuity restriction, and this allows for solving the problem by a marching procedure in which each computation at a certain node requires only the values from neighboring nodes in analogy with an explicit difference method. Since the stability conditions associated with explicit methods allow for typical stepsizes to be chosen having the same order in time and space, the method will be very effective.

For a brief presentation, we choose another PDE in one space dimension and time:

$$\frac{\partial u}{\partial t} + \frac{\partial f(u)}{\partial x} = 0 \,,$$

where f may be nonlinear. Just as with the Galerkin method, the equation is multiplied by a function $\phi(x)$ and integrated. However, in this case the integration is done over a subinterval $I_j = [x_{j-1/2}, x_{j+1/2}]$ of the computational domain. After integration-by-parts, the result is

$$\int_{I_j} \frac{\partial u}{\partial t} \phi \, dx + f\big(u(x_{j+1/2})\big)\phi(x_{j+1/2}) - f\big(u(x_{j-1/2})\big)\phi(x_{j-1/2}) - \int_{I_j} f(u)\frac{d\phi}{dx} \, dx = 0 \,.$$

Continuity of the solution at the endpoints is given up, but there has to be some connection to the next subinterval. This is achieved by substituting the "flux function" $f(u)$ by a numerical flux function $g(u^-, u^+)$ with

$$u^-(x) = \lim_{\delta \to 0} u(x - \delta), \quad \delta > 0 \,,$$

$$u^+(x) = \lim_{\delta \to 0} u(x + \delta), \quad \delta > 0 \,,$$

i.e., $g(u^-, u^+) = g(u, u) = f(u)$ if u is continuous. We have discussed such numerical flux functions in Sect. 4.3.5 above.

The piecewise polynomials form a finite dimensional space $\mathscr{S}_h$. The numerical solution v is then defined in the following compact way:
* Find the function $v \in \mathscr{S}_h$ such that

$$\int_{I_j} \frac{\partial v}{\partial t} \phi \, dx + g\big(v^-(x_{j+1/2}), v^+(x_{j+1/2})\big)\phi(x_{j+1/2})$$

$$- g\big(v^-(x_{j-1/2}), v^+(x_{j-1/2})\big)\phi(x_{j-1/2}) - \int_{I_j} f(v)\frac{d\phi}{dx} \, dx = 0 \,,$$

is satisfied for all j and all $\phi \in \mathscr{S}_h$.

It remains to choose the function $g(v^-, v^+)$ and the form of v on each interval. For the simple equation $\partial u/\partial t + \partial u/\partial x = 0$ with characteristics going from left to right, we choose g as

$$g(u^-, u^+) = u^-,$$

and v as a piecewise linear function with

$$v(x, t) = a_j(t)\phi_j(x) + b_j(t)\psi_j(x), \qquad x \in I_j,$$

where

$$\phi_j = \frac{1}{\Delta x_j}(x_{j+1/2} - x), \qquad \psi_j = \frac{1}{\Delta x_j}(x - x_{j-1/2}), \qquad x \in I_j,$$

are the basis functions. The integrals are now easy to evaluate, and the final equations for the coefficients are

$$\frac{\Delta x}{3}\frac{da_j}{dt} + \frac{\Delta x}{6}\frac{db_j}{dt} = b_{j-1} - \frac{1}{2}b_j,$$

$$\frac{\Delta x}{6}\frac{da_j}{dt} + \frac{\Delta x}{3}\frac{db_j}{dt} = \frac{1}{2}a_j - \frac{1}{2}b_j, \qquad j = 1, 2, \ldots, N. \tag{5.12}$$

The PDE requires a boundary condition at the left, hence b_0 can be considered as known. The unknowns da_j/dt, db_j can easily be solved for by solving a 2 by 2 system.

So what is new with this method compared to the regular Galerkin method? If the coefficients a_j, b_j are ordered as vectors, the left-hand side can also be seen here as a mass matrix M multiplying these vectors. However, the important point is the block-diagonal structure, i.e., there is no coupling between the coefficients in different subintervals. and this allows for an explicit difference method in time.

The generalization to higher order accuracy is obtained by using higher order polynomials on each subinterval. It results in larger size blocks in the mass matrix, but these block systems can be solved for one-by-one as in the linear case. The generalization to more dimensions is obtained by going to triangles, etc., just as for traditional finite elements.

As we have demonstrated here, this type of method has a certain advantage for a hyperbolic PDE, where there is a natural marching direction. There were various similar developments for elliptic PDEs in the 1970s, one of them being the article [4] by Baker. He used the concept *"nonconforming elements"*; other authors used *"discontinuous elements"*. Nowadays these methods go under the name *"Discontinuous Galerkin methods"*.

Stability and accuracy of these methods do not follow as easily as for Galerkin methods, and this has been a very active area of research during the last decades. This theory is outside the scope of this book, but most of it is found in the book [90] by Hesthaven and Warburton.

5.6.5   Grid Generation

By a structured grid we mean either uniformly distributed grid-points, or alternatively, a point-distribution that is obtained by a smooth transformation of a uniform grid. An unstructured grid consists of points that are located without any particular structure. Unlike finite difference methods, finite element methods can easily be applied to such grids. This is a fundamental strength of FEM, since the geometry of the computational domain may be irregular, and furthermore, the solutions may have an irregular behavior. However, the construction of unstructured grids is certainly nontrivial in two- and higher dimensions, and this becomes a challenge by itself. Even if a distribution of points is defined in the whole domain in some way characterizing the variation of the solution, the question is how these points should be connected to each other so that a complete set of subdomains is obtained. We have seen that finite elements are well suited to be defined on triangles in two dimensions. Indeed, the problem of forming a set of triangles from a set of points was solved by Boris Delaunay (1890–1980) in 1934; the method goes under the name Delaunay triangulation, see [40]. Delaunay was Russian, and his French surname goes back to his ancestor De Launay who was a French army officer captured during the Napoleon invasion in 1812.

Delaunay was a student of Georgy Voronoi, and the triangulation method is based on Voronoi diagrams as shown in Fig. 5.10. The green points are given, and the blue convex Voronoi polygons are such that they contain exactly one generating point and such that every point in the polygon is closer to its generating point than any other. The sides of the triangle are then defined as the red lines perpendicular to the polygon sides and connecting the corresponding point pairs. (The case with more than two points on a straight line must be treated separately.) The Delaunay triangulation can be generalized to more than two dimensions. It was of course not

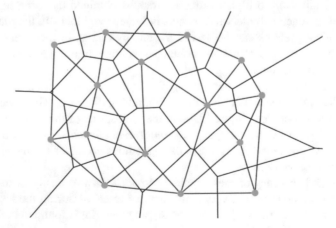

Fig. 5.10 Voronoi diagram (blue) and triangulation (red) (David Coeurjolly, http://liris.cnrs.fr/~dcoeurjo/teaching/ENS2017/~images/voronoi-delaunay.png)

intended for generating a finite element grid, but it is one of the most frequently used methods in up-to-date FEM-solvers.

Another frequently used principle is the advancing front method. It was introduced in 1985 by S.H. Lo, see [118], and followed up in 1987 by Peraire et al., see [133]. In two dimensions the main idea is that nodes connected by straight lines are first introduced on the outer boundary. This forms an initial "front", and if there are inner boundaries as well, the front consists of several parts. Triangles are then generated one-by-one with the pieces on the front forming one side of each triangle. A new front is obtained, and the process continues using the same principle. Special procedures must be applied when different parts of the front intersect, so that a final triangulation is obtained.

For effective FEM solvers, the generation of grids is a significant part of the solution process. Many variants of generating techniques have been developed, most of them based on one or both principles described above, but sometimes requiring some manual efforts in support of the automatic procedure. There are also many variants of automatic or semi-automatic adaptive techniques for improving the grid so that it becomes better suited to represent variations of the solution.

5.7   The Fast Fourier Transform

5.7.1   The Algorithm

In Sect. 3.6 we discussed the discrete Fourier transform (DFT) and its wide areas of application. We repeat basic formulae here, but change the normalization constant $1/\sqrt{2\pi}$ so it agrees with the convention in most of the literature. If a set of N numbers $\{f_j\}$ are given, the DFT is defined by

$$\tilde{f}_\omega = \sum_{j=0}^{N-1} f_j e^{-i\omega x_j}, \qquad \omega = 0, 1, \ldots, N-1. \tag{5.13}$$

The given numbers f_j are recovered by the inverse DFT:

$$f_j = \frac{1}{N} \sum_{\omega=0}^{N-1} \tilde{f}_\omega e^{i\omega x_j}, \qquad j = 0, 1, \ldots, N-1. \tag{5.14}$$

The Fourier transform formulas are well defined for any sequence of numbers $\{f_j\}$. However, if the sequence represents N uniformly distributed point-values $f(x_j)$ of a 2π-periodic function $f(x)$ such that $x_j = jh$ and $f_{j+N} = f_j$, the function $g(x)$

defined by

$$g(x) = \frac{1}{N} \sum_{\omega=0}^{N-1} \tilde{f}_\omega e^{i\omega x}, \qquad 0 \le x \le 2\pi$$

is an interpolating function that approximates $f(x)$. This is the basis for the fundamental importance of the DFT. However, direct computation of the sums require $\mathcal{O}(N^2)$ arithmetic operations, and if the DFT is a central part in an inner loop of a certain algorithm, the computation might be way too demanding for large data sets.

Gauss (and later others) already knew how to speed up the straightforward algorithm, but the real breakthrough came when the Fast Fourier Transform (FFT) was presented, analyzed and implemented in the article [24] published by James W. Cooley and John W. Tukey in 1965. The FFT can be constructed for any integer N, but the case where N is a power of two is the most well known since the computational gain then is maximal. The basic idea is to partition the computation into two smaller parts:

$$\tilde{f}_\omega = \sum_{j=0}^{N/2-1} f_{2j} e^{-i\omega x_{2j}} + e^{-i\omega h} \sum_{j=0}^{N/2-1} f_{2j+1} e^{-i\omega x_{2j}} = S_\omega + e^{-i\omega h} R_\omega,$$

$$\omega = 0, 1, \dots, N-1.$$

By periodicity

$$S_{\omega+N/2} = S_\omega, \qquad R_{\omega+N/2} = R_\omega,$$

and we have

$$\tilde{f}_\omega = \begin{cases} S_\omega + e^{-i\omega h} R_\omega, & 0 \le \omega \le N/2 - 1, \\ S_{\omega-N/2} + e^{-i\omega h} R_{\omega-N/2}, & N/2 \le \omega \le N - 1. \end{cases}$$

Furthermore,

$$e^{-i(\omega+N/2)h} = e^{-i\omega h - i\pi} = e^{-i\omega h} e^{-i\pi} = -e^{-i\omega h},$$

which allows for reducing the computation of $e^{-i\omega h}$ by a factor of two as well:

$$\tilde{f}_\omega = \begin{cases} S_\omega + e^{-i\omega h} R_\omega, & 0 \le \omega \le N/2 - 1, \\ S_{\omega-N/2} - e^{-i(\omega+N/2)h} R_{\omega-N/2}, & N/2 \le \omega \le N - 1. \end{cases}$$

The computation is now broken down into two DFT of half the length. If the work for the full DFT is cN^2 for some constant c, then the work for a DFT of half the length is $cN^2/4$. This reduction process can now be repeated once more and then continued until each piece contains only two values.

The total work w_m for computing the FFT can now be estimated. For $N = 2^m$, the remaining number of arithmetic operations when $\{R_\omega\}$ and $\{S_\omega\}$ are computed is at most $2 \cdot 2^m$. We then have

$$w_m \leq 2w_{m-1} + 2 \cdot 2^{m-1},$$

and it follows by induction that

$$w_m \leq 2m \cdot 2^m = 2N \log_2 N.$$

This reduction of the work is even more pronounced in several dimensions, and explains why the Fourier transform is such an effective tool in scientific computing.

If N is not a power of two, the principle of partitioning the computation can be applied anyway. Any factorization of N can be used to reduce the work, but we shall not go further into this here.

After the Cooley–Tukey article was published, new investigations were initiated to look back into the history to see if the same idea had been around before, and sure enough, Gauss' name came up again. Clearly, Gauss did not publish any manuscript on the topic while he was alive, but after his death, a presentation of his collected works [64] was published in 1866, and a method for the fast Fourier transform is included there. In 1977 Goldstine made an investigation into the historical facts, see [72], but the most careful research was presented by Heideman et al., see [87]. The fact that Gauss wrote in Latin, and that he used quite different notation and formulation of the problem makes it difficult to compare to more recent publications. However their conclusion is that Gauss' method is basically equivalent to the FFT as we know it. However, he worked with the trigonometric series formulated in terms of real functions as one cos-series and one sin-series.

The FFT algorithm is certainly one of the most important in the history of computational mathematics. The main reason is that it reduces the operation count from $\mathcal{O}(N^2)$ to $\mathcal{O}(N \log N)$, where N is the number of given points. In higher dimensions the gain is even more pronounced since the count $\mathcal{O}(N^{2d})$ reduces to $\mathcal{O}(dN^d \log N)$. Furthermore, the discrete Fourier transform has such a high number of applications apart from the obvious signal processing applications. We have seen that new methods for solving PDE problems were developed just because the computational work could be reduced to roughly the same work as for traditional methods, but with a much higher accuracy.

Various lists of the most important algorithms are often presented, and of course there are different opinions about which ones should be there. However, almost every list includes FFT. Actually, it has by some authors been considered as the most important algorithm of all time, for example by Gilbert Strang, see [152].

In the next section we shall present one of the many applications, where the impact of the FFT certainly has revolutionized medical tools based on X-ray investigations.

5.7.2   An Application of FFT: Computed Tomography

The problem of finding out what is hiding inside a living person is as old as the existence of man. The breakthrough came with the invention of the X-ray machine; Fig. 5.11 shows an X-ray image of human lungs. However, the image shows only a *projection* of the chest. We know the location and geometry of the lungs quite well, but the image by itself doesn't show anything about the variation in the direction perpendicular to the image surface. The image represents variations in the density of the tissue, but in principle these variations could be located anywhere between the front and back part of the chest. The revolutionary new tool was obtained by the invention of computed tomography. The idea is to take a series of two-dimensional X-rays in the x/y-plane by moving the machine (or the object) in the z-direction. The challenge is to produce a digital image from each of the X-ray projections; this is the central part of the algorithm and has to be done fast. These images are then merged together so that a three-dimensional image is obtained.

Figure 5.12 shows an X-ray beam with the energy E_0 passing through an object hitting the detector with the energy E_1. If $f(s)$ denotes the density of the material,

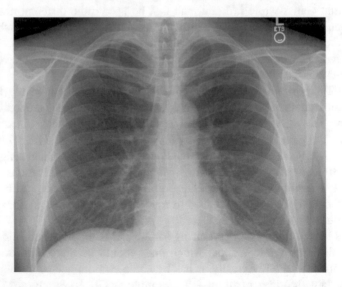

Fig. 5.11   X-ray image of the chest (©Zephyr/Science Photo Library 12303355)

Fig. 5.12 X-ray beam

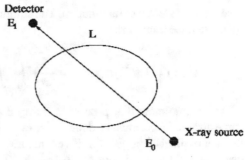

Fig. 5.13 X-ray beams at angle θ

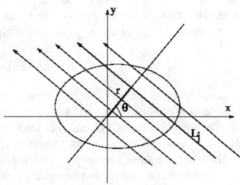

we have

$$E_1 = E_0 e^{-\int_L f(s)ds},$$

i.e.,

$$\int_L f(s)ds = ln(\frac{E_0}{E_1}) = p,$$

where p is measured. The problem at hand is to find $f(s)$, but this is of course not possible, since many different density distributions give the same integral. However, if many X-ray beams are sent through the object in parallel from positions all the way around the object, as in Fig. 5.13, the density can be found as a unique distribution. The function p is now a function of r and θ, where r is the distance from the origin to a given ray as illustrated in the figure. By symmetry, it is sufficient to vary the angle θ between 0 and π. The connection between p and f was found by the Austrian mathematician Johann Radon (1887–1956), and is expressed through what is now called the *Radon transform* defined by an integral

$$p(r, \theta) = [\mathcal{R}f](r, \theta) = \int_{-\infty}^{\infty} \int_{-\infty}^{\infty} f(x, y)\delta(x cos\theta + y sin\theta - r)dxdy,$$

where δ is the Dirac delta function. The solution to this integral equation is obtained by the inverse transform

$$f(x, y) = [\mathscr{R}^{-1}p](x, y) = \frac{1}{2\pi^2} \int_0^\pi \int_{-\infty}^\infty \frac{\frac{\partial p}{\partial r}(r, \theta)}{x\cos\theta + y\sin\theta - r} dr d\theta.$$

The requirement that p be known for all r and θ is of course not fulfilled and some discretization must be done. Since every digital image is represented by a finite number of pixels (x_j, y_k), this is very natural. Furthermore, the detectors register the energy at a finite number of points (r_j, θ_k). Therefore the discrete Radon transform is represented by a matrix A, and with $\mathbf{f}$ and $\mathbf{p}$ denoting the vectors containing the point values, we get an algebraic system of equations

$$A\mathbf{f} = \mathbf{p}. \tag{5.15}$$

One can show that the inverse Radon transform is ill-posed in the sense that it in general it produces solutions with nonphysical perturbations of high frequency, i.e., the image becomes blurred. This property is also present for the corresponding discrete operator A^{-1}, and extra smoothing is required.

However, the fundamental problem of reconstructing the image is solved. The scanner with the necessary computer built in was constructed by the British engineer Godfrey Hounsfield (1919–2004). The first version was complete and functioning in 1971, and it was described in 1973, see [93]. At this time the South African physicist Allen McLeod Cormack (1924–1998) had already worked out essentially the same mathematics independently; it was published in 1963 and in 1964, see [25] and [26]. He was using the Fourier transform, but at this time the FFT didn't exist. His articles didn't get much attention at the time, but they certainly did when he and Hounsfield shared the Nobel Prize in Physiology or Medicine in 1979.

The system (5.15) is enormous since good image quality requires a large number of pixels. Therefore straightforward Gauss elimination is not a possible alternative in practice. Various iterative methods have been developed, but the algebraic reconstruction technique has been replaced by other techniques based on the Fourier transform. We shall describe here one of these versions called the *Direct Fourier Method* (DFM).

Assuming that the object is finite and located inside the square $[-c, c] \times [-c, c]$, the projection is located in the rectangle $[-d, d] \times [0, \pi)$, where $d \geq c\sqrt{2}$. Then

$$\hat{p}(\rho, \theta) = \frac{1}{\sqrt{2\pi}} \int_{-d}^d p(r, \theta)e^{-2\pi i\rho r} dr \tag{5.16}$$

is the one-dimensional Fourier transformation of p in the r-direction, and we have

$$p(r, \theta) = \frac{1}{\sqrt{2\pi}} \int_{-\infty}^\infty \hat{p}(\rho, \theta)e^{2\pi i\rho r} d\rho.$$

The density function f can be represented as

$$f(x, y) = \int_{-\infty}^{\infty} \int_{-\infty}^{\infty} \hat{f}(\xi, \eta) e^{2\pi i(\xi x + \eta y)} d\xi \, d\eta, \tag{5.17}$$

where $\hat{f}$ is the two-dimensional Fourier transform

$$\hat{f}(\xi, \eta) = \int_{-c}^{c} \int_{-c}^{c} f(x, y) e^{-2\pi i(\xi x + \eta y)} dx \, dy.$$

The basis for the reconstruction is the *Slice Theorem* derived by R.N. Bracewell in 1956, see [11]:

$$\hat{p}(\rho, \theta) = \hat{f}(\rho \cos \theta, \rho \sin \theta), \qquad -\infty < \rho < \infty, \quad 0 \le \theta < \pi. \tag{5.18}$$

By changing variables from (ξ, η) to (ρ, θ) in (5.17), we get

$$f(x, y) = \int_{0}^{\pi} \int_{-\infty}^{\infty} \rho \hat{p}(\rho, \theta) e^{2\pi i(x \cos \theta + y \sin \theta)\rho} d\rho \, d\theta.$$

It is now clear how to construct the algorithm. Given p, we first compute $\hat{p}$ by using the fast Fourier transform FFT for the discrete version of (5.16), then use the slice theorem (5.18), and finally compute f using the FFT for the discrete version of (5.17).

There are two remaining difficulties. The FFT gives $\hat{p}$ at a uniform distribution of points in the ρ-direction, but we need $\hat{f}$ at uniformly distributed points in the variables ξ and η. These points don't coincide, and some sort of interpolation is required. Standard polynomial interpolation is no good as shown in [77], where another method based on Fourier series expansion is suggested.

The other difficulty is the well known Gibbs phenomenon. A typical image contains sharp contours, one of them being the outer boundary of the object, which means that the density function f has discontinuities, and as demonstrated in Sect. 3.6.4, they cause erroneous oscillations in the Fourier approximation. Smoothing operators may be applied, but the effect is that even the contours are smoothed out to a certain degree, and this goes against the efforts to get sharp images.

A trick without this side effect is the following. Let f_0 be a simple function containing the major discontinuity, and with the corresponding Radon transform $\mathcal{R}(f)$. The procedure is then based on the formula

$$f = \mathcal{R}^{-1}(p - p_0) + f_0. \tag{5.19}$$

If f_0 is simple enough, there is an analytical form of p_0, which is subtracted from the original p, and the numerical approximation corresponding to (5.19) is computed.

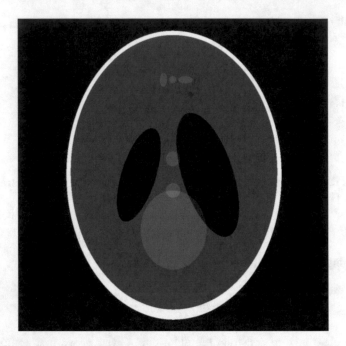

Fig. 5.14 Shepp-Logan phantom image

In the tomography community, the Shepp–Logan phantom image is quite common as a test case. In that case contours are defined mathematically, which makes it easy to find p_0 analytically. Figure 5.14 shows the result from a computation with the direct Fourier method applied on the square domain with 512×512 pixels, which gives a quite good quality.

The direct method has been described here in order to show the use of the Fourier transform in a simple way. Another method, actually used in most tomography machines, is the *Filtered Backprojection Method (FBM)*. It is also based on the Fourier transform and has a filter built in for taking care of the Gibbs phenomenon. The final formula is obtained in polar coordinates and reads as

$$f(r, \theta) = \int_0^\pi \int_{-\infty}^\infty \left(\int_{-\infty}^\infty p(\xi, \phi) e^{2i\rho\xi i} d\xi \right) |\rho| e^{2\pi\rho r \cos(\theta-\phi)i} d\rho \, d\phi.$$

The integrals over the variables ξ and ρ are computed in their discrete form by using the FFT. The algorithm is very fast and when the machine has finished the X-ray projections, the resulting image is computed within seconds even in the true three-dimensional case.

Computed tomography has had an enormous impact on the investigation of a large number of illnesses in which information about the inside of the body is required, but also on many other applications where tomography is used. The real breakthrough was the construction of the tomography machine with the reconstruction of the image built in, which also resulted in a Nobel Prize. However, the later development of applying the discrete Fourier transform was also a kind of breakthrough, since it speeded up the algorithm considerably thanks to the FFT algorithm.

5.8 Spectral Methods

Spectral methods can be used as a name for all methods that approximate the solution of a certain problem as a combination of given basis functions

$$u(\mathbf{x}) = \sum_j c_j \phi_j(\mathbf{x}),$$

and the challenge is to find the coefficients c_j. If the problem is to simply represent a set of measurements of a given function, the procedure is called spectral analysis. However, the concept of spectral methods usually refers to methods that use series expansions to represent solutions of differential equations. This technique has been used by applied mathematicians for a long time, and in Sect. 3.6 we discussed an example involving a simple form of the heat equation and the use of Fourier series. For more general differential equations, it is not possible to find the coefficients c_j by analytical means, and we have to develop numerical methods using a finite number of modes ϕ_j.

5.8.1 Pseudo-Spectral Methods

The first serious attempt with spectral methods was done by Steve Orszag (1943–2011) in the late 1960s. He was at the mathematics department at MIT, but was interested in numerical simulation of fluid dynamics and turbulence. He used the incompressible Navier–Stokes equations as a model, and Fourier expansions for the solution. The equations are nonlinear, and the multiplication of two Fourier series is the source of computational difficulties. He published the first article [125] in 1969, and followed up with several others in the following years.

In 1970 he was visiting the National Center for Atmospheric Research (NCAR) in Boulder, Colorado, where he conducted large scale computations on the powerful computer that was located there. At the same time Heinz Kreiss visited NCAR where he met Joseph Oliger at the computational group. They began looking into this new type of method from a theoretical point of view, and analyzed it for simple model problems. Their work and Orszag's resulted in a new version called *pseudospectral methods*, which was the name used by Orszag in his articles [126] published in 1971 and [127] published in 1972. Kreiss and Oliger, who published their first paper [102] on the topic in 1972, called it the Fourier method, since it dealt with periodic solutions and the discrete Fourier transform. We shall now show how it is constructed.

The FFT is the central tool, but it is convenient to change the notation from Sect. 5.7 slightly and assume that N is an even number such that

$$x_j = jh, \qquad j = 0, 1, \ldots, N, \ (N+1)h = 2\pi.$$

Furthermore, we reintroduce the normalizing factor $1/\sqrt{2\pi}$ such that the transform and its inverse get a symmetric form. Then the discrete Fourier transform is defined by

$$\tilde{u}_\omega = \frac{1}{\sqrt{2\pi}} \sum_{j=0}^{N} u_j e^{-i\omega x_j} h, \qquad \omega = -N/2, -N/2+1, \ldots, N/2,$$

with the inverse

$$u_j = \frac{1}{\sqrt{2\pi}} \sum_{\omega=-N/2}^{N/2} \tilde{u}_\omega e^{i\omega x_j}, \qquad j = 0, 1, \ldots, N.$$

The interpolating function

$$v(x) = \frac{1}{\sqrt{2\pi}} \sum_{\omega=-N/2}^{N/2} \tilde{u}_\omega e^{i\omega x}$$

can now be differentiated with respect to x:

$$w(x) = \frac{dv}{dx} = \frac{1}{\sqrt{2\pi}} \sum_{\omega=-N/2}^{N/2} i\omega \tilde{u}_\omega e^{i\omega x}.$$

The grid-values w_j of the new function are obtained by a three-step procedure using the FFT:

1. Fourier transform the set of grid-values u_j.
2. Multiply the Fourier coefficients $\tilde{u}_\omega$ by $i\omega$.
3. Use the inverse FFT for computing the grid-values w_j.

The grid-values of u and w form two vectors

$$\mathbf{u} = \begin{bmatrix} u_0 \\ u_1 \\ \vdots \\ u_N \end{bmatrix}, \qquad \mathbf{w} = \begin{bmatrix} w_0 \\ w_1 \\ \vdots \\ w_N \end{bmatrix},$$

and the relation between them can be written as

$$\mathbf{w} = S\mathbf{u},$$

where S is an $(N + 1) \times (N + 1)$ matrix. From this it follows that the eigenvalues and eigenvectors of S are

$$S \begin{bmatrix} 1 \\ e^{i\omega h} \\ \vdots \\ e^{i\omega Nh} \end{bmatrix} = i\omega \begin{bmatrix} 1 \\ e^{i\omega h} \\ \vdots \\ e^{i\omega Nh} \end{bmatrix}, \qquad \omega = -N/2, -N/2 + 1, \ldots, N/2.$$

Consider now the linear PDE in one space dimension

$$\frac{\partial u}{\partial t} + a(x, t)\frac{\partial u}{\partial x} = 0$$

with 2π-periodic solutions in x. Traditional difference methods replace the differential operator $\partial/\partial x$ with a finite difference operator. The Fourier method replaces it with the operator S that we have just constructed. When discretizing only in the x-direction, the grid-values u_j depend on t, which means that also the Fourier coefficients $\tilde{u}_\omega$ depend on t. With the vector-matrix notation above, we can write the semidiscrete system as

$$\frac{d\mathbf{u}}{dt} + AS\mathbf{u} = 0, \qquad A = diag\big(a_0(t), a_1(t), \ldots, a_N(t)\big).$$

It remains to discretize in time, and this is done by a regular difference method. For a hyperbolic PDE as this one, the leap-frog scheme can be used, see Sect. 3.7.2. Since the largest eigenvalue of S in magnitude is $N/2$, the stability condition for constant coefficients a is

$$\frac{a\Delta t}{h} \leq \frac{1}{\pi},$$

i.e., π times stricter than with the second order finite difference operator D_0. However, the accuracy is so much better, at least for smooth solutions, which makes the method superior when it can be applied. Requiring periodic solutions limits of course the application field considerably.

We note that the Fourier method can be applied equally easy for nonlinear problems of the type where $a = a(u)$ as well. The pseudo-spectral character of the method avoids the problem that arises when multiplying two finite Fourier expansions with each other. However, a straightforward application of Fourier methods does not work well when discontinuities and shocks are present, due to the Gibbs phenomenon as discussed in Sect. 3.6.4. Some sort of filtering must be introduced, and one example is described for the tomography problem in Sect. 5.7.2.

The Fourier method can be seen as a special type of *collocation method*. For this class of methods, the solution is approximated by a function expansion in which the coefficients are determined such that the resulting function satisfies the differential equation exactly at a number of "collocation points". For the problem above, the Fourier expansion satisfies the PDE exactly at the grid-points x_j, or rather the lines (x_j, t), before discretization in time. The principle had been used for ODE much earlier, among others by C. Lanczos, and he refers to the collocation points as "selected points", see [105] published in 1938.

For initial boundary value problems with nonperiodic solutions on a finite domain, the basic Fourier method cannot be applied. However, methods of the pseudo-spectral type can still be used if the basis functions are chosen differently; Orszag already in 1972 uses Chebyshev polynomial expansions, see [127]:

$$u(x) = \sum_{n=0}^{N} c_n T_n(x).$$

We described these polynomials in Sect. 3.6 together with their recursive definition in the interval $-1 \le x \le 1$. The relation

$$T_n(\cos \theta) = \cos(n\theta), \qquad n = 0, 1, \ldots$$

makes it possible to compute the coefficients c_n for the interpolating polynomial by a cosine transform. However, the grid-points must then be chosen as

$$x_j = \cos \frac{j\pi}{N}, \qquad j = 0, 1, \ldots, N,$$

which are located much closer to each other near the boundaries rather than in the middle of the interval. This might be an advantage for problems with boundary layers with sharp gradients near the boundary, but in general it is a disadvantage.

Other types of classical polynomials, as for example Legendre polynomials, have been used as well, and over the following decades the theory and practical applicability was advanced considerably, also for the more traditional techniques based on Galerkin and tau-approximations. David Gottlieb (1944–2008) was one of the leading researchers in this area, with the article [76] as an example of his stability theory for Chebyshev pseudo-spectral methods. Together with Jan Hesthaven and Sigal Gottlieb he wrote the book [91], which contains most of the essential material on the topic.

5.8.2 Spectral Element Methods

Finite element methods as developed during the first decades of their existence, were based on piecewise polynomials of low order as basis functions. In fluid dynamics with complicated flow patterns, the degree was usually at most two. At the other extreme were the spectral methods with order of accuracy being of the same order as the number of interpolation points in each direction. However, the limitation on the geometry introduced by the requirements on the location of the interpolation points was severe for general problems. Anthony Patera came up with the idea of combining the two methods, and he introduced the *spectral element method*, see [130]. He used the new method to solve the incompressible Navier–Stokes equations

$$\frac{\partial \mathbf{v}}{\partial t} + (\mathbf{v} \cdot \nabla)\mathbf{v} = -\nabla p + \nu \nabla^2 \mathbf{v},$$

$$\nabla \cdot \mathbf{v} = 0,$$

where $\mathbf{v}$ and p are the velocity vector and the pressure respectively, and ∇ is the gradient vector. As a simple model problem, he used the one-dimensional equation

$$u_t + u_x = \nu u_{xx}$$

with a combination of the second order Adams–Bashforth and the trapezoidal rule for time discretization. The idea was to divide the computational domain into sub-domains $\{\Omega_k\}$, and then introduce a spectral representation of the solution in each of them. Patera chose Chebyshev polynomials with local interpolation points,

$$x_j^{(k)} = \cos \frac{j\pi}{N^{(k)}}, \qquad j = 0, 1, \ldots, N^{(k)},$$

where $N^{(k)}$ is the number of points in Ω_k. This allowed for efficient computation since the fast cos-transform could be used as described above.

Multidomain spectral methods had been developed earlier, but the spectral element method is based on the variational formulation which provides a general treatment of the internal boundaries between elements.

The spectral element method quickly gained interest, which resulted in theoretical error estimates and applications to various problems in which sufficient accuracy had been difficult to achieve. However, one should be aware that there were still restrictions on the geometry as a consequence of the limitation to rectilinear elements for spectral representations.

5.9   Multiscale Methods

Many processes and states in physics and engineering are characterized by variations on different scales. The movement of water waves are usually considered at the
scale of meters called macroscale, but there is also variation on a much smaller scale
at the surface; we call this the microscale. Still it makes sense to model the water
waves on the macro level without including the microscale surface effects. However,
in order to get very accurate results, even the microscales should be included in
the model, but then the computation becomes too heavy if we are interested in
waves over a large geographical domain. Some special technique must be introduced
to make the computation reasonable with respect to the size. Since there is an
interaction between the two scales, it is a real challenge to construct methods that
takes this into account without solving the whole system on the microscale.

There is also another type of multiscale technique that can be applied to a model
that is formulated without explicit separation of scales. The numerical method is
constructed so that it takes the presence of varying scales into account. If the smaller
scales are limited to certain subdomains, the computational grid can be constructed
with extra refinement in these domains. Another example is multigrid methods that
work on different refinement levels as part of the solution procedure. Still another
example concerns multipole methods (5.9.2), in which the effect on the solution
from some variables can be approximated in a simplified way corresponding to a
macroscale. These two methods have had a strong effect on computations during the
last decades, and we will discuss them in the following sections. Then we follow up
with recent developments for the first type of problems, in which different scales are
present as parameters in the equations.

5.9.1   Multigrid Methods

The development of computational methods was very fast in the 1950s. The
dominating problem class was the solution of partial differential equations in a
large variety of application areas. Finite difference methods that were used at
the time led to the need for solving very large systems of algebraic equations.
For elliptic boundary value problems these systems are unavoidable; for implicit
methods applied to time-dependent problems they occur as well. Direct methods
could not be used, since the storage requirements widely exceeded the available
internal computer memory. Iterative methods were easy to apply since the systems
have a very simple and sparse structure, but the convergence rate was way too slow,
causing great frustration.

In the Soviet Union at the time, computers were not that well developed
compared to what was available in the west. Furthermore, the most powerful were
used for military purposes, and scientists at the universities and other research
institutes could not count much on them. But the Steklov Mathematical Institute

in Moscow gradually turned to more nonmilitary projects, and one of these was weather prediction. In the simple model that was used, the time-dependent equations includes the Poisson equation in space, which has to be solved for each timestep. This is the dominating part of the computation, and requires a very fast Poisson solver. Otherwise the computer simulation of the weather would be far slower than the real time process, and therefore useless.

One of the mathematicians at the Steklov Institute who had to deal with this was Radii Petrovich Fedorenko (1930–2010). He used a simple iterative method, and became very frustrated by the slow convergence. He really wanted to understand the reasons for this, and studied the process in detail. We shall give an elementary example in order to explain what he found.

The boundary value problem

$$-\frac{d^2u}{dx^2} + au = F(x), \qquad 0 \le x \le 1,$$

$$u(0) = 0,$$

$$u(1) = 0,$$

is approximated by the standard three-point difference approximation

$$-u_{j-1} + 2u_j - u_{j+1} + ah^2u_j = h^2 F(x_j), \qquad j = 1, 2, \ldots, N,$$

$$u_0 = 0, \tag{5.20}$$

$$u_{N+1} = 0,$$

where h is the stepsize. With the notation $\mathbf{u}$ and $\mathbf{b}$ for the vectors with elements u_j and $b_j = h^2 F(x_j)$ respectively, the system of equations is

$$A\mathbf{u} = \mathbf{b},$$

where A is an $N \times N$ matrix. The system is solved by the damped Jacobi method

$$\mathbf{u}^{(n+1)} = \mathbf{u}^{(n)} - \theta D^{-1}(A\mathbf{u}^{(n)} - \mathbf{b}), \qquad n = 0, 1, \ldots,$$

where θ is a parameter with $0 \le \theta \le 1$. Here we consider the special case $a = 0$ giving the diagonal $D = 2I$ in the matrix A. The convergence properties are determined by the eigenvalues λ_j of the iteration matrix $M = I - \theta D^{-1}A$, and they are

$$\lambda_k = 1 - 2\theta \sin^2 \frac{k\pi h}{2}, \qquad k = 1, 2, \ldots, N.$$

The corresponding eigenvectors have the elements

$$\sin(k\pi hj), \qquad j = 1, 2, \ldots, N,$$

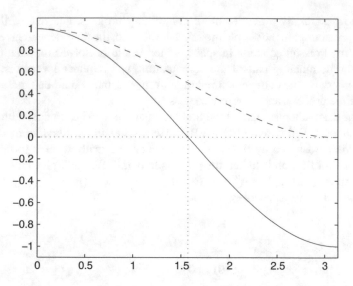

Fig. 5.15 Eigenvalues $\lambda(\xi)$ of the damped Jacobi method, $\theta = 1$ (solid line), $\theta = 0.5$ (dashed line)

and for each wave number k, we can see them as grid functions for increasing integers j. These functions oscillate faster for larger k, and for $k = N$, the function is very close to $(-1)^j$. For a given h, it is convenient to study eigenvalues as functions of $\xi = k\pi h$, where $0 < \xi < \pi$. Figure 5.15 shows the eigenvalues

$$\lambda(\xi) = 1 - 2\theta \sin^2 \frac{\xi}{2}, \qquad 0 < \xi < \pi$$

as functions of ξ for $\theta = 1$ and $\theta = 0.5$. With $\theta = 0.5$, we have

$$\max_{\pi/2 \leq \xi \leq \pi} |\lambda(\xi)| = 0.5,$$

i.e., the convergence properties are very good for high wave numbers. However, for any choice of θ, the convergence is very poor for lower wavenumbers. Fedorenko got a brilliant idea when he realized this. A certain large ξ_0 with good convergence properties corresponds to the wavenumber $k_0 = \xi_0/(\pi h)$ on the given grid. On a grid twice as coarse the corresponding wavenumber $k_1 = \xi_0/(\pi 2h)$ is half as big. Hence, by somehow transferring the problem to a coarser grid, the ξ-interval with good convergence properties corresponds to a lower wave number part of the solution.

We now introduce the label h as a subscript for the grid G and as a superscript for the corresponding vectors and matrices. To make it possible to alternate between the fine grid G_h and the coarse grid G_{2h}, there is a need to transfer grid functions between them. This can be done in many ways; here we choose

$$u_j^{2h} = \frac{1}{4}(u_{2j-1}^h + 2u_{2j}^h + u_{2j+1}^h), \qquad j = 0, 1, \ldots, (N+1)/2,$$

where it is assumed that $N+1$ is an even number. This *restriction* can be expressed by the vector/matrix form

$$\mathbf{u}^{2h} = R\mathbf{u}^h, \qquad R = \frac{1}{4}\begin{bmatrix} 2 & 1 & & & \\ 1 & 2 & 1 & & \\ & & \ddots & & \\ & & & 1 & 2 \end{bmatrix}$$

with the boundary points excluded. Figure 5.16 illustrates this procedure together with the *prolongation* in the other direction:

$$\mathbf{u}^h = P\mathbf{u}^{2h}, \qquad P = R^T = \frac{1}{2}\begin{bmatrix} 2 & & & \\ 1 & 1 & & \\ & 2 & & \\ & 1 & & \\ & & \ddots & \\ & & & 1 \\ & & & 2 \end{bmatrix}.$$

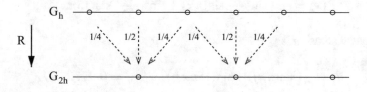

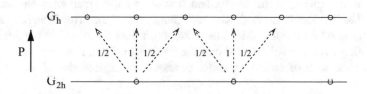

Fig. 5.16 Restriction to a coarser grid and prolongation to a finer grid

The simplest version of the multigrid method is the two-grid method. Let $\mathbf{v}$ be an approximate solution, and $\mathbf{r} = \mathbf{b} - A\mathbf{v}$ the corresponding residual. We have

$$\mathbf{v} = A^{-1}(\mathbf{b} - \mathbf{r}),$$

hence the true solution can be defined by

$$A\mathbf{w} = \mathbf{r},$$

$$\mathbf{u} = \mathbf{v} + \mathbf{w}.$$

One step of the two-grid algorithm is

1. Define an initial guess $\mathbf{u}^{h(0)}$.
2. Compute a better approximation $\mathbf{v}^h$ by one or more iterations on the original fine grid G_h.
3. Compute the residual $\mathbf{r}^h = \mathbf{b}^h - A^h\mathbf{v}^h$.
4. Compute the residual $\mathbf{r}^{2h} = R\mathbf{r}^h$ by restriction.
5. Solve $A^{2h}\mathbf{w}^{2h} = \mathbf{r}^{2h}$ by one or more iterations on the coarse grid G_{2h}.
6. Compute the new approximation by $\mathbf{u}^{h(1)} = \mathbf{v}^h + P\mathbf{w}^{2h}$.

In the general multigrid method, one continues at step 5 above by going to a still coarser grid G_{4h} and applying the two-grid procedure once more. If the grid is such that $N + 1 = p2^q$ with integers p and q, we have $q + 1$ possible levels. If p is small enough, one can solve the final system by a direct method on the coarsest grid.

With $h = 1/16$ and

$$F(x) = -2\left(\sin^2\frac{\pi}{32}\sin(\pi x) + \sin^2\frac{15\pi}{32}\sin(15\pi x)\right),$$

the numerical solution is

$$u_j = \frac{1}{2}\left(\sin(\pi x_j) + \sin(15\pi x_j)\right), \qquad j = 0, 1, \ldots, 17.$$

The two different wave numbers are clearly seen in Fig. 5.17.

The dotted line in the right part of the figure shows the error $u_j - u(x_j)$ for the zero initial guess, while the dashed line shows the error after one iteration with the damped Jacobi method. The oscillatory part of the error is already almost completely gone on the fine grid. The solution on the coarse grid is obtained by a direct method, and the solid line shows the final error. Even if the example is very simple for the sake of clearness, it illustrates the great potential of the multigrid method.

Fedorenko published his introduction of the two-grid method in 1964, see [52]. The method was based on the convergence analysis for iterative methods that he had presented earlier, see [51]. However, the paper and the method didn't get the attention it deserved at the time, probably to a large extent depending on the lack

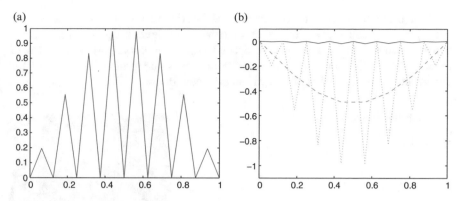

Fig. 5.17 Exact solution and error with the two-grid method. (**a**) Exact solution. (**b**) Error

of interest in Russian journals in the west. But things would change, and that was due to Achi Brandt (1938–), an Israeli scientist at the Weizmann Institute in Rehovot, Israel. Few people have had such a strong impact as he had on the development of a particular method. He gave a talk on what he called multi-level adaptive techniques at a conference on numerical methods in fluid mechanics in Paris in 1972, which resulted in the publication [12]. That was the beginning of a long series of talks worldwide over many years, in which he demonstrated the efficiency of the method for increasingly more complicated problems. The article [13] contains many different aspects of the method, including programming and storage details. The theoretical convergence analysis is based on Fourier analysis, just as we demonstrated for the simple example above.

Achi Brandt (Courtesy Achi Brandt)

Wolfgang Hackbusch (1948–) at the University of Kiel in Germany was another who picked up the multigrid idea in the 1970s. Fourier analysis gives very good indications for the convergence behavior of the method, but since the solutions to most problems cannot be represented as Fourier series, it is not enough for strictly general convergence theorems. Hackbusch bridged this gap and gave strict convergence proofs for a number of special cases. His book [84] is loaded with theoretical results, not only for linear boundary value problems, but also for eigenvalue problems and nonlinear differential equations.

Wolfgang Hackbusch
(Courtesy Wolfgang
Hackbusch)

Multigrid methods are today a standard tool for solving large systems of equations. We recall that a general direct solver based on Gaussian elimination has an operation count of $\mathcal{O}(N^3)$, where N is the number of unknowns. Under quite general conditions, it is possible to construct a multigrid method that converges to a prescribed accuracy with a total number cN of arithmetic operations. Real world problems often have several millions of unknowns, and the difference in computing time and storage requirements is enormous. No solver can do better than this asymptotically, the only question is the size of the constant c.

Several advanced generalizations have been made; one of them is algebraic multigrid methods. Even if "grid" is still part of the name, the main point is that these methods are not based on any geometrical grid. Instead, one starts with the system of equations in matrix/vector form, and constructs the different levels by extracting subsets of unknowns without any reference to geometry.

One warning is suitable here. There is no black box multigrid solver that guarantees a successful converged computation of the solution to any given system of equations. And even if a converged solution is obtained, it may take a much longer than expected. Some applications are more difficult than others, and require being handled in a certain way, including careful tuning of certain parameters.

5.9.2   Fast Multipole Methods (FMM)

The solution of many problems in computational mathematics can be formulated as computation of a sum

$$u(x) = \sum_{k=1}^{N} w_k K(x, y_k), \tag{5.21}$$

where u is to be evaluated at N points x_j, $j = 1, 2, \ldots, N$. In some cases this sum is exact, as in the N-body problem describing the interaction between N different bodies located at N different locations. It can also be an approximation of an integral

$$u(x) = \int_{0}^{Y} w(y) K(x, y)\, dy,$$

as for the simplest form of the heat conduction. Obviously in direct evaluation of (5.21) requires $\mathcal{O}(N^2)$ arithmetic operations, which may be way too high for many realistic applications. A smart summation algorithm that reduces the work significantly has high strategic value.

We have already seen an example of such an algorithm already, namely the fast Fourier transform. This is a very special case where $K(x_j, y_k) = e^{2\pi i j k / N}$, and it was shown in Sect. 5.7 how the work can be reduced to $\mathcal{O}(N \log N)$ algebraic operations. This algorithm uses the fact that the points x_j are uniformly distributed and that the function $e^{2\pi i y}$ describes the unit circle for real y. For general $K(x, y)$ and arbitrarily located points x_j, the challenge is much greater.

Vladimir Rokhlin, born in Russia in 1952, got his Ph.D. at Rice University, Texas. He was interested in fast numerical methods for integral equations and published the paper [141] on this topic in 1985, when he got his first position at Yale University. There he worked together with Leslie Greengard who got his Ph.D. there in 1987 (actually, he got an MD there the same year as well). They published the much celebrated paper [79] the same year, and the Fast Multipole Method was born.

We will not present details of the method here, but in summary it goes like this. For illustration we choose the electrostatics problem, which is very similar to the n-body problem. In two dimensions we use the complex variable $z = x + iy$ to represent the location (x, y) and we work with complex potentials $\phi(z)$. First, we consider the potential at points far away from a charge. It can be shown that a single point charge of strength q at z_0 induces the potential

$$\phi_{z_0}(z) = q \left(\log z - \sum_{k=1}^{\infty} \frac{1}{k} \left(\frac{z_0}{z} \right)^k \right)$$

at any point z with $|z| > |z_0|$. The main idea of FMM is to treat charges that are located close to each other as if they are one charge when it comes to the influence on the potential far away. Assume that the charges q_j, $j = 1, 2, \ldots, m$, are located at z_j, $j = 1, 2, \ldots, m$, with $|z_j| < r$. Then, for any z with $|z| > r$, the potential is

$$\phi(z) = \sum_{j=1}^{m} q_j \log z - \sum_{k=1}^{\infty} \frac{1}{kz^k} \sum_{j=1}^{m} q_j z_j^k.$$

This multipole expansion contains an infinite number of terms, and a truncation is required. With $\phi(z, p)$ denoting the approximate potential with p terms in the sum, the following error bound holds:

$$|\phi(z) - \phi(z, p)| \leq \frac{1}{(p+1)(1 - |r/z|)} \left| \frac{r}{z} \right|^{p+1} \sum_{j=1}^{m} |q_j|.$$

When computing the total potential at a certain point z_0, the charges are partitioned into clusters where some are near z_0 and some are far away. As an example, we consider a square with a nonuniform charge distribution. One way is to construct adaptive solutions so that each subsquare contains a small number of charges, as shown in Fig. 5.18. The potential is computed for each point, using the correct formulas for all charges located in the same box and for the boxes having at least one boundary point in common. For all other points, multipole expansion is used.

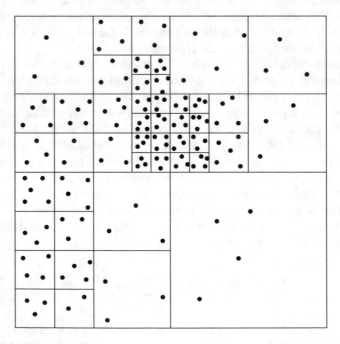

Fig. 5.18 Adaptive partitioning

Many more details are required to implement the algorithm, for example the transformation to local expansions around the center of each box. Furthermore, there are two different strategies, one leading to an $\mathcal{O}(N \log N)$ algorithm, the other of order $\mathcal{O}(N)$. And, of course, it is the problem in three dimensions that is of prime interest. In that case the expansions are expressed with spherical harmonics as basis functions. For a complete description of the algorithm, we recommend the report [6].

5.9.3 Heterogeneous Multiscale Methods (HMM)

Heterogeneous multiscale methods are based on a very recent technique which was first introduced by Weinan E and Björn Engquist in 2003, see [161]. It is designed for models that contain different scales explicitly in the differential equation. For simplicity, we illustrate this technique using a system of ordinary differential equations that has the form

$$\begin{aligned}\frac{du}{dt} &= -\frac{1}{\varepsilon}\big(u - f(v)\big),\\[2mm]\frac{dv}{dt} &= g(u, v),\end{aligned} \tag{5.22}$$

where ε is a small positive constant. In reality one could easily solve the full system with a very small timestep corresponding to the microscale of order ε, but here we assume that this requires too much computing power, and we try to solve it with significantly less work. Given a value $u(t_0)$ at a certain point $t = t_0$ in time, the negative large coefficient multiplying u in the first equation takes u quickly close to a slowly varying quantity $u = f(v)$, and we get the macroscale equation

$$\frac{dv}{dt} = g\big(f(v), v\big) =: G(v). \tag{5.23}$$

For simplicity, we use the forward Euler method for integrating the system, and with the macroscale timestep h, we get

$$v_{n+1} = v_n + h\tilde{G}_n(v_n), \tag{5.24}$$

where $\tilde{G}$ is an approximation of the forcing function G. The question is how $\tilde{G}$ should be defined. With u_n and v_n known, the system is decoupled and a new u is obtained by freezing $v = v_n$ at the current level. The microscale equation for u in (5.22) is solved using a small timestep k:

$$u_{n,m+1} = u_{n,m} - \frac{k}{\varepsilon}\big(u_{n,m} - f(v_n)\big), \qquad m = 0, 1, \ldots, M. \tag{5.25}$$

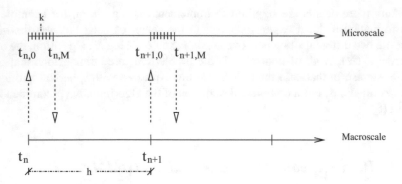

Fig. 5.19 The HMM procedure

The number of steps M is chosen so that u has converged sufficiently close to the almost stationary state. The new value of u is now used to define

$$\tilde{G}(v_n) = g(u_{n,M},\ v_n),$$

and (5.24) can then be advanced one step on the macroscale (Fig. 5.19).

The amplification factor for (5.25) is

$$Q = 1 - \frac{k}{\varepsilon},$$

and stability requires that $k \leq \varepsilon$. So, what is gained? The timestep k is of microscale order, and this is what we wanted to avoid. But here is the key to the efficiency of the method. If

$$k = \alpha\varepsilon, \qquad 0 < \alpha < 1,$$

microsteps converge to the correct solution quickly, and M is independent of ε. This can be compared to the properties of the differential equation itself. For the simple case $f = 0$, the homogeneous equation has the solution

$$u(t) = ce^{-t/\varepsilon}$$

and $|u(t)| \leq \delta$ for $t = \mathcal{O}(\varepsilon)$. Even if the stepsize is $\mathcal{O}(\varepsilon)$, the necessary time interval for integration is of the same order. On the discrete side, this means that the number of steps stays bounded, independent of ε.

The example treated here is special in the sense that the microscale equation is stiff and converges fast to a function with macroscale variation. But HMM can be applied to many other types of equations as well. A common case is when one variable has a quasi-periodic behavior, i.e., there is a periodic behavior on the macroscale, but within each period, there is a microscale variation. In this case, the simple introduction of the fast variable $u_{n,M}$ into the macroscale equation is replaced by introducing an average of all the values

$$u_{n,m}, \qquad m = 0, 1, \ldots, M.$$

HMM has been further developed and analyzed during the last years, and is used extensively for various applications. A survey covering the time until 2012 is found in [1].

5.10   Singular Value Decomposition (SVD)

In Sect. 4.4.3 we discussed computing the eigenvalues λ_j of a square $m \times m$-matrix A. The similarity transformation

$$S^{-1}AS = \Lambda = \mathrm{diag}(\lambda_1, \lambda_2, \ldots, \lambda_m)$$

is called an eigenvalue decomposition of A. However, such a decomposition does not always exist. A corresponding singular value decomposition is closely connected to the eigenvalue concept, but is less restrictive. Not only does it always exist for any square matrix, but also for rectangular $m \times n$-matrices, where $m > n$. If A is real, then

$$U^T A V = \Sigma$$

is a singular value decomposition, where the matrices U and V are both orthogonal, i.e., $U^T U = I$ and $V^T V = I$. Unfortunately, there are notational differences in the literature when defining the SVD. Here we define U and V as square matrices with m and n columns respectively. This means that Σ is an $m \times n$ matrix with non-negative elements and the form

$$\Sigma = \begin{bmatrix} \sigma_1 & & & \\ & \sigma_2 & & \\ & & \ddots & \\ & & & \sigma_n \\ & & & \\ & & & \end{bmatrix}, \tag{5.26}$$

where empty positions mean zeros. However, there is also another convention used, where U is an $m \times n$ matrix, in which case

$$
\Sigma = \begin{bmatrix} \sigma_1 & & & \\ & \sigma_2 & & \\ & & \ddots & \\ & & & \sigma_n \end{bmatrix} \tag{5.27}
$$

is a square $n \times n$ matrix. Since $A^T = V \Sigma U^T$, we have $A^T A = V \Sigma^2 V^T$, showing that the singular values are the square roots of the eigenvalues of $A^T A$. This is sometimes used as the definition of singular values.

Note that even if we consider the case $m = n$, the singular values are in general not the same as the eigenvalues, since the requirement of orthogonal matrices U and V means that A must be symmetric and positive definite in such a case where $U = V$. So, what is the practical use of an SVD?

The SVD can be traced back to Gauss (which should not be a surprise at this point). He was concerned with the method of least squares, and we shall see later how it is connected to the SVD. However, the first published fundamental work on the existence and construction of the SVD is due to Euginio Beltrami (1835–1899) and Camille Jordan (1838–1921). An excellent survey of the early history of the SVD is presented by G.W. Stewart in [151].

Before going further, we introduce the *pseudo-inverse* A^+ of a matrix A. As the name indicates, it is a concept that may be of help when the true inverse A^{-1} does not exist, and it is also defined for rectangular matrices. There are different ways of defining the pseudo-inverse, but usually one has the Moore–Penrose pseudo-inverse in mind. It got its name from the American mathematician Eliakim Hastings Moore (1862–1932) and the British mathematician and physicist Roger Penrose (1931–), see [121] and [132]. The formal definition for an $m \times n$ matrix A is an $n \times m$ matrix A^+ satisfying the following four conditions:

$$
AA^+A = A,
$$

$$
A^+AA^+ = A^+,
$$

$$
(AA^+)^T = AA^+,
$$

$$
(A^+A)^T = A^+A.
$$

The pseudo-inverse always exists and is unique. Obviously it equals the true inverse when the latter exists. Furthermore, if A has full column rank, then there is a simple explicit formula

$$
A^+ = (A^T A)^{-1} A^T,
$$

which is the left inverse, i.e., $A^+ A = I$. Similarly, if A has full row rank, then

$$A^+ = A^T (A A^T)^{-1},$$

which is the right inverse, i.e., $A A^+ = I$.

The most common use of the pseudo-inverse is probably when solving singular least square problems. As an example we consider the over-determined linear system of equations

$$A \mathbf{x} = \mathbf{b},$$

where $\mathbf{x}$ and $\mathbf{b}$ have n and m elements respectively. The solution to this problem by using the least square method was discussed in Sect. 3.4, but there it was assumed that A has full column rank n such that the normal equations with $A^T A$ as the coefficient matrix have a unique solution. According to the formulas above, this solution is

$$\mathbf{x} = A^+ \mathbf{b}.$$

The beautiful fact is that even if the column rank is less than n, in which the least square problem has no solution at all, or it has many solutions, the pseudo-inverse always provides a solution, and it can be shown that with $\mathbf{z} = A^+ \mathbf{b}$ and with $\| \cdot \|$ denoting the Euclidean norm, we then have

$$\| A \mathbf{z} - \mathbf{b} \| = \min_{\mathbf{x}} \| A \mathbf{x} - \mathbf{b} \|.$$

Furthermore, in the case with many solutions $\mathbf{x}$ to the least square problem, the pseudo-inverse captures the one with the shortest length.

Now to the connection between the SVD and the pseudo-inverse. The pseudo-inverse of Σ in (5.26) can easily be obtained by the definition. If all singular values are nonzero, then

$$\Sigma^+ = \begin{bmatrix} 1/\sigma_1 & & & \\ & 1/\sigma_2 & & \\ & & \ddots & \\ & & & 1/\sigma_n \end{bmatrix}.$$

If a certain σ_j is zero, then $1/\sigma_j$ is replaced with zero. With this definition, it can be shown that the pseudo-inverse for the original matrix A is

$$A^+ = V \Sigma^+ U^T.$$

This formula is the basis for computing the pseudo-inverse, and we need a numerical method for computing the SVD, i.e., the singular values σ_j and the matrices U and V.

Gene Golub (1932–2007) had the key role in this development. He is definitely one of the most well known numerical analysts in linear algebra. He got his Ph.D. in 1959 at the University of Illinois at Urbana-Champaign, and arrived at Stanford University in 1962 where he stayed throughout his career until his death. Just about every numerical analyst visiting Stanford some time during the period 1965–2007 have enjoyed Golub's hospitality in his house on the Stanford campus.

Gene Golub   (Author:
Konrad Jacobs; Source:
Oberwolfach Photo
Collection. Copyright MFO)

The first complete numerical method for SVD was presented in 1965 by Golub and Kahan in [73]. It was refined in 1970, see [74], and we shall here summarize how it works. The straightforward approach would be to use the fact noted above that the eigenvalues of the $n \times n$-matrix $A^T A$ are $\sigma_1^2, \sigma_2^2, \ldots, \sigma_n^2$. Hence, if these are found, the singular values are obtained by taking the square roots. But Golub and Reinsch show that there is reason to do better than applying a straightforward eigenvalue algorithm to $A^T A$. Instead, A is reduced to an upper bidiagonal form by two sequences of Householder transformations $P^{(j)}$, $Q^{(j)}$ such that

$$
P^{(n)} \cdots P^{(1)} A Q^{(1)} \cdots Q^{(n-2)} =
\begin{bmatrix}
q_1 & e_2 & & & \\
& q_2 & e_3 & & \\
& & \ddots & \ddots & \\
& & & q_{n-1} & e_n \\
& & & & q_n
\end{bmatrix},
$$

where empty positions mean zeros. This matrix J has the same singular values as A, and with

$$
G^T J H = \Sigma,
$$

we have

$$G^T P^T J Q H = \Sigma, \qquad P = P^{(1)} \cdots P^{(n)}, \quad Q = Q^{(1)} \cdots Q^{(n-2)}.$$

(Here Σ has the form (5.27).) The final step is to construct the SVD of a bidiagonal matrix. This is done through a special procedure based on successive Givens rotations; for details we refer the reader to [74]. The characteristic feature of the algorithm is that these transformations are applied directly to the matrix J, giving singular values without making the detour via the eigenvalues of $J^T J$.

Most of the existing software packages of today for SVD and pseudo-inverses are based on this algorithm, even if a few other variants have been developed.

Finally we note that singular values are well defined also for complex matrices A by defining U and V as unitary matrices, i.e., $U^*U = I$ and $V^*V = I$.

5.11   Krylov Space Methods for Linear Systems of Equations

In 1951 the Swiss mathematician Eduard Stiefel (1909–1978) visited the National Bureau of Standards, Boulder, USA, where he met Magnus Hestenes (1906–1991) who had a position at UCLA, Los Angeles. It turned out that they had very similar ideas about a new approach to solving linear systems of equations. They wrote the joint paper [89], which later became the basis for a new class of methods that is perhaps the dominating one today. We begin by describing their method.

We want to solve the system

$$A\mathbf{x} = \mathbf{b}, \tag{5.28}$$

in which we make the restriction that A is a symmetric positive definite matrix. This system is closely connected to the quadratic form

$$f(\mathbf{x}) = \frac{1}{2}\mathbf{x}^T A \mathbf{x} - \mathbf{b}^T \mathbf{x},$$

since the unique minimum value of $f(\mathbf{x})$ is attained at $\mathbf{x} = A^{-1}\mathbf{b}$. Therefore, we can convert the problem of solving (5.28) to the problem of minimizing $f(\mathbf{x})$. Optimization had been a topic for many mathematicians and other researchers over the years, since there were so many applications leading to such problems. However, at the time there was essentially one numerical method for practical problems, and it was called the method of steepest descent. Thinking of a two-dimensional problem and with $f(\mathbf{x})$ representing a surface in three-dimensional space, this means that standing at a certain point $\mathbf{x}^{(0)}$, one continues searching for the next point in the direction $\mathbf{d}^{(0)}$ where the surface has the steepest slope. This direction is given by $-\mathbf{grad}\, f(\mathbf{x}^{(0)})$. The minimum $f(\mathbf{x}^{(1)})$ in that direction is found, and the iterative procedure continues with a search in the direction that is pointing along the steepest slope at the new point $\mathbf{x}^{(1)}$. The problem with the method is that it converges very slowly for general problems, since the local properties near a given point do not tell much about the global properties. Hestenes and Stiefel decided to do something about this.

Two vectors $\mathbf{x}$, $\mathbf{y}$ are orthogonal if they satisfy $\mathbf{x}^T \mathbf{y} = 0$. In the steepest descent method, two consecutive search directions $\mathbf{d}^{(n)}$, $\mathbf{d}^{(n+1)}$ are orthogonal. The Hestenes–Stiefel idea was to generalize the concept of orthogonal vectors to conjugate vectors, which satisfy $\mathbf{x}^T A \mathbf{y} = 0$, and then use conjugate directions in an iterative procedure. Assume that the latest search direction is $\mathbf{d}^{(n-1)}$ leading to the new point $\mathbf{x}^{(n)}$ and the residual

$$\mathbf{r}^{(n)} = \mathbf{b} - A\mathbf{x}^{(n)} .$$

We make the ansatz

$$\mathbf{d}^{(n)} = \mathbf{r}^{(n)} + \beta \mathbf{d}^{(n-1)},$$

which gives the condition

$$(\mathbf{d}^{(n-1)})^T A (\mathbf{r}^{(n)} + \beta \mathbf{d}^{(n-1)}) = 0,$$

and the solution is

$$\beta = -\frac{(\mathbf{d}^{(n-1)})^T A \mathbf{r}^{(n)}}{(\mathbf{d}^{(n-1)})^T A \mathbf{d}^{(n-1)}} .$$

If all directions are chosen this way, it can be shown that they all become mutually conjugate.

The new point is to be found along the new direction as

$$\mathbf{x}^{(n+1)} = \mathbf{x}^{(n)} + \alpha \mathbf{d}^{(n)},$$

where the parameter α is chosen so that $f(\mathbf{x})$ is minimal. The new point, the new residual, and the new search direction are

$$\mathbf{x}^{(n+1)} = \mathbf{x}^{(n)} - \frac{(\mathbf{d}^{(n)})^T \mathbf{r}^{(n)}}{(\mathbf{d}^{(n)})^T A \mathbf{d}^{(n)}} \mathbf{d}^{(n)} ,$$

$$\mathbf{r}^{(n+1)} = \mathbf{b} - A\mathbf{x}^{(n+1)} ,$$

$$\mathbf{d}^{(n+1)} = \mathbf{r}^{(n+1)} - \frac{(\mathbf{d}^{(n)})^T A \mathbf{r}^{(n+1)}}{(\mathbf{d}^{(n)})^T A \mathbf{d}^{(n)}} \mathbf{d}^{(n)} .$$

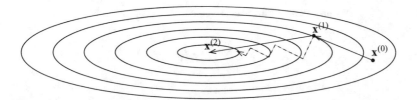

Fig. 5.20 The conjugate gradient method in two dimensions and steepest descent

This is the conjugate direction method (or CG-method). Figure 5.20 shows the behavior of the CG-method and the steepest descent method for a two-dimensional problem.

When the method is applied to an $N \times N$ system, it turns out that it produces the correct answer after exactly N steps, at least if N is of moderate size. At first glance this might seem surprising, but it should not be. It is quite easy to show that the residuals are mutually orthogonal, i.e.,

$$(\mathbf{r}^{(j)})^T \mathbf{r}^{(k)} = 0, \qquad j \neq k.$$

We are working in an N-dimensional space here, and there can be no more than N nonzero vectors

$$\mathbf{r}^{(j)}, \qquad j = 0, 1, \ldots, N - 1$$

that are orthogonal. As a consequence $\mathbf{r}^{(N)} = 0$, which shows that $\mathbf{x}^{(N)}$ is the correct solution.

Each step of the conjugate gradient method requires three matrix/vector multiplications, so the full algorithm takes $\mathcal{O}(N^3)$ a.o., which is the same as Gaussian elimination. The question then arises: have we gained anything compared to a direct solver? Yes, we have. A direct solver has to be completed to the end before a solution is defined, while the CG-method produces an answer $\mathbf{x}^{(j)}$ at each step and can be considered as an iterative method. And it turns out that the convergence is quite good as long as the system is not too ill conditioned.

Many researchers picked up the CG-method and tried it on various problems during the 1950s. Even if it was considered as a direct method, it was noted that quite accurate solutions were obtained in much less time than N iterations. On the other hand, it was also noted that for some problems, one had to iterate more than N steps to get an acceptable solution. And even worse, sometimes the algorithm didn't converge at all. So what was the problem?

The algorithm uses multiplications of vectors by the matrix A again and again. That makes it natural to consider the Krylov space $\mathcal{K}_n$ defined as

$$\mathcal{K}_n = span\{\mathbf{r}^{(0)}, A\mathbf{r}^{(0)}, A^2\mathbf{r}^{(0)}, \ldots, A^{n-1}\mathbf{r}^{(0)}\},$$

which means that any vector $\mathbf{r}$ in this space can be expressed as

$$\mathbf{r} = \sum_{j=0}^{n-1} c_j A^j \mathbf{r}^{(0)},$$

where c_j are constants. It is not difficult to show that the space spanned by the first n directions $\mathbf{d}^{(j)}$, $j = 0, 1, \ldots, n-1$, of the CG-method is exactly the Krylov space $\mathcal{K}_n$, as defined above. The method is therefore called a Krylov space method.

The space is named after the Russian mathematician and engineer Alexei Krylov (1863–1945), who had quite an interesting career. He entered the Naval College at the age of 15 and continued at the Naval Academy of Saint Petersburg, from which he graduated in 1890 and then continued working as a teacher and scientist. He did significant research in the rolling theory of ships and many other problems connected to shipbuilding. He became the Chief Executive Officer of the Russian Society for Shipbuilding and Trade and after the revolution he transferred the whole merchant fleet of the society to the Soviet government. As a scientist he was very active at an older age, and in fact he published the article [103], which is of interest here, in 1931.

It is hard to find land areas that are named after famous mathematicians. However, the Krylov Peninsula in Antarctica is one such name. On the other hand, there are quite a few on the moon since many mathematicians could also be considered as astronomers. On the back side of the moon, there are many Russian names, and the reason for this is that the Soviet Union sent the space probe Lunar 3 around the moon and photographed the back side which no one had ever seen before. The Krylov Crater is one example.

The year 1952 was an amazing year when it comes to solving linear systems of equations. Three people had very similar ideas about a new method. Two of them, Stiefel and Hestenes, met at the National Bureau of Standards, worked out the CG-method and published it jointly. The third one, Cornelius Lanczos (1893–1974.), (see Sect. 5.8) worked there as well, and he published his article [106] the same year. The Lanczos method was formulated as a Krylov space method from the very beginning. Furthermore, the article dealt with the eigenvalue problem as well, which was solved using the same Krylov space principle.

With the Krylov space machinery, one can prove a number of estimates concerning convergence rate. Each iteration adds a new dimension to the approximating space. With the initial guess $\mathbf{x}^{(0)}$ approximating the true solution $\mathbf{x}^*$, it is interesting to know at what stage the ratio $\|\mathbf{x}^{(n)} - \mathbf{x}^*\|/\|\mathbf{x}^{(0)} - \mathbf{x}^*\|$ is smaller than a given tolerance ε. The answer is that the number of iterations n must be chosen such that

$$n \geq \frac{1}{2}\sqrt{\text{cond}(A)}\,\log\frac{2}{\varepsilon},$$

where $\text{cond}(A)$ is the condition number

$$\text{cond}(A) = \|A\| \cdot \|A^{-1}\|.$$

For symmetric positive definite matrices A, we have

$$\text{cond}(A) = \frac{\lambda_{\max}}{\lambda_{\min}},$$

where λ denotes the eigenvalues of A. Another estimate is

$$\|\mathbf{x}^{(n)} - \mathbf{x}^*\|_A \le 2 \left(\frac{\sqrt{\text{cond}(A)} - 1}{\sqrt{\text{cond}(A)} + 1} \right)^n \|\mathbf{x}^{(0)} - \mathbf{x}^*\|_A,$$

Alexei Nikolaevich Krylov
(Structurae; International
Database for Civil and
Structural Engineering, Image
No. 221394; Public domain)

where

$$\|\mathbf{x}\|_A = \left(\mathbf{x}^T A \mathbf{x} \right)^{1/2}.$$

No matter which estimate we choose, it is obvious that the condition number of A is crucial for the convergence. If the system is stiff, i.e., if the matrix A has a wide spectrum of eigenvalues, we can expect trouble.

Everything that is said so far is based on the assumption that all computations are carried out exactly. This is of course impossible in practice, since rounding errors always enter. The question then is: can the algorithm stand small perturbations, or, in other words, is the algorithm stable? In fact the CG-method is not, which eventually was found to be the cause of much trouble. A thorough analysis of this problem was presented in 1959 in the Swiss publication [47] by Engeli et al. The basic reason for the trouble is that with a given fixed vector $\mathbf{x}$, the vectors $A^j \mathbf{x}$, $j = 0, 1, \ldots$, become almost linearly dependent for large j. This means that they are not good

basis vectors for the space, which in turn is the reason for the initially surprising
fact that convergence in at most N iterations is not always achieved.

Cornelius Lanczos
(University of Manchester,
Cornelius Lanczos)

Various kinds of stabilization methods have been devised, and the method has
been generalized to general systems of equations. One of these generalizations is the
Generalized Minimal Residual Method (GMRES) developed by Saad and Schultz
and published in 1986, see [146].

As we have seen above, the condition number of the matrix plays a fundamental
role for the convergence properties, and if the problem could be transformed
resulting in a new and better conditioned matrix, the situation would improve. Such
a procedure is called preconditioning, and can be seen as a special case of a very
general concept in everyday life, meaning that an object is given a certain treatment
in advance, such that some process applied to it becomes more effective. If one
is going to run a marathon race, it is a good idea to improve the condition of the
body prior to the race, otherwise the body will not perform well when needed. In
computational mathematics, it means that a problem is transformed into a different
form before a certain numerical algorithm is applied to it. For linear systems of
equations, it means that a transformation is applied to it before an iterative method
is used. If M is a matrix, the original system

$$A\mathbf{x} = \mathbf{b}$$

is premultiplied by M^{-1}, resulting in a new system

$$M^{-1}A\mathbf{x} = M^{-1}\mathbf{b}$$

with the same solution $\mathbf{x}$. The idea is that the new form allows for a faster
convergence. For the CG-method described above, we need the condition number
cond($M^{-1}A$) to be significantly smaller than cond(A). The choice $M = A$ is of
course perfect when it comes to minimizing the condition number, but since we
need to solve systems with the original coefficient matrix in order to find M^{-1}, we

have achieved nothing. The algorithm requires computation of the type $\mathbf{z} = M^{-1}A\mathbf{r}$ at each step. This is a two stage procedure;

$$\mathbf{y} = A\mathbf{r},$$
$$M\mathbf{z} = \mathbf{y}.$$

(5.29)

If A is sparse with a total of $\mathcal{O}(N)$ nonzero elements, the first part is an $\mathcal{O}(N)$ procedure, and if the algorithm for solving the second part can be kept at this order as well, we are in good shape. One such case is obtained with an incomplete factorization. When performing a factorization $A = LR$, with lower and upper triangular matrices on the right-hand side, there is a fill-in effect, such that these matrices are much denser than A. There are various ways of keeping L and R sparse as well, one way being to neglect all terms below a certain threshold ε. This results in an incomplete factorization $\tilde{A} = \tilde{L}\tilde{R}$, and we choose $M = \tilde{A}$. In this way, solving the second equation in (5.29) is obtained by two back substitutions with a total of $\mathcal{O}(N)$ operations.

Another approach is to use one type of iterative method as a preconditioner to another iterative method. For example, the multigrid method can be used as a preconditioner to the CG-method. The connection to the matrix M in this case is that the multigrid method can be seen as producing the exact solution to a system with M as coefficient matrix.

Where can the origin of the preconditioning concept be found? Jacobi had already experimented with orthogonal transformations to change the matrix, so that it would become diagonally dominant which is a requirement for convergence of the Jacobi method, see [94]. In 1937, the Italian mathematician Lamberto Cesari used a polynomial $P(A)$ as preconditioner to reduce the condition number of A, see [19]. Preconditioners for CG-methods were discussed in 1959, see [47] (already mentioned above).

Krylov space methods have had an enormous expansion over the years, and they now exist in many different versions with many different preconditioners. There is no way to measure what kind of numerical method is most frequently used for a certain mathematical problem, but when it comes to general linear systems of equations, a good guess is that the class of Krylov space methods is a strong candidate for the top position. A good survey of iterative methods and preconditioners is found in [7].

Finally we note that for a general system of equations, in which nothing is known about its eigenvalue distribution or other specific properties, there is still no iterative method, preconditioned or not that guarantees fast convergence.

5.12   Domain Decomposition

The solution of boundary value problems for elliptic PDEs took a significant step forward when the British mathematician George Green (1793–1841) constructed the Green function, which made it possible to write down the solution in an explicit form. For example, let Ω be a domain in the two-dimensional space with boundary Γ, and consider the problem

$$Lu = F(x, y), \qquad (x, y) \in \Omega,$$
$$u = 0, \qquad (x, y) \in \Gamma,$$

where L is an elliptic differential operator. Then the solution can be written in the form

$$u(x, y) = -\int\int_{\Omega} G(x, y, \xi, \eta) F(\xi, \eta) d\xi d\eta,$$

where $G(x, y, \xi, \eta)$ is the Green function. For certain problems, such as the Poisson equation and simple geometries, G is known explicitly. The solution u can then be computed at any point in the domain by evaluating the integral.

The question is now whether these explicit forms of the solution can be used for more general geometries. Hermann Schwarz (1843–1921) came up with an idea how this could be achieved, and he published it in 1870, see [150].

Hermann Schwarz
(Photographer unknown.
Source: http://www-gap.dcs.
st-and.ac.uk/~history/
PictDisplay/Schwarz.html.
Wikimedia*)

As a simple 2D-example, consider a domain composed of a rectangle and a circle as shown in Fig. 5.21. The boundary of the full domain is $\Gamma_1 \cup \Gamma_2$. The dashed contours indicate a full rectangular domain Ω_1 and a full circular disc Ω_2 that overlap each other. Assuming that the problem can be solved on each of these subdomains, the full problem is solved by iteration, alternating between Ω_1 and

Fig. 5.21 A domain composed of a rectangle and a circle

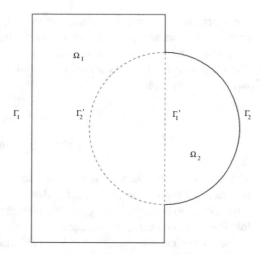

Ω_2. Starting from an initial guess $\mathbf{u}^0 = [u_1^0 \ u_2^0]$ with u_1^0 in Ω_1 and u_2^0 in Ω_2, the algorithm is

1. *Find the solution u_1^{n+1} in Ω_1 with u_2^n as boundary values on Γ_1'.*
2. *Find the solution u_2^{n+1} in Ω_2 with u_1^{n+1} as boundary values on Γ_2'.*

The method is called the Schwarz Alternating Method, and it can be extended to problems with domains that are composed of any finite number of subdomains. There are several convergence proofs for it; for second-order elliptic PDE, convergence was proven by Mikhlin, see [120]. Besides certain regularity conditions, the most essential condition is that the overlapping domains have a positive fixed area.

Schwarz didn't have any computer; he had to assume that each of the subdomain problems could be solved by analytical means. Little did he know that his principle of domain decomposition would come to new life a century later. That was the time when high performance computers were constructed as a collection of several processors P_j that operate in parallel, and the partition into subdomains opened up for parallel algorithms. However, the Schwarz alternating algorithm above cannot use the advantage of parallel treatment of the subdomains. The reason is that the second stage of the algorithm has to wait for the first stage to be completed, since the boundary data u_1^{n+1} are needed for the computation. The processor P_2 is idle while P_1 works, and vice versa. The cure is obvious, leading to the algorithm

1. *Find the solution u_1^{n+1} in Ω_1 with u_2^n as boundary values on Γ_1'.*
2. *Find the solution u_2^{n+1} in Ω_2 with u_1^n as boundary values on Γ_2'.*

This version is called the Schwarz parallel algorithm, since it allows for parallel processing of the two subdomains. However, in this particular case with only two subdomains, one doesn't gain anything. A simple check shows that every second step produces the same solutions as does the alternating algorithm for each step, and nothing is gained. Since the alternating algorithm uses later information on the boundary, there is no contradiction in this phenomenon. However, in the general case where there are more than two subdomains, there is a gain.

The parallel algorithm was introduced in 1988 by Pierre-Louis Lions, see [115]. The case with an arbitrary number of subdomains Ω_j, $j = 1, 2, \ldots, N$, can be written in the simple form

$$Lu_j^{n+1} = F \quad \text{in } \Omega_j,$$

$$u_j^{n+1} = u_k^n \quad \text{on } \Gamma_{jk}.$$

The choice of boundaries Γ_{jk} in which to pick the Dirichlet boundary data from for each subdomain is not trivially defined. The reason for this is the obvious requirement to construct a single-valued continuous solution u by putting the pieces together. We omit these details here.

For numerical computation the Schwarz methods must be discretized, primarily by finite difference methods or finite element methods. The resulting methods fall in the more general category of domain decomposition methods. Interestingly, the introduction of parallel computers was not the only driving force for these methods. During the 1970s there was a rapid development of fast Poisson solvers. In the beginning they were constructed for the original Poisson equation $\Delta u = F$, where Δ is the Laplace operator, but later for more general elliptic equations. The crucial point is that the geometry is simple, such as a rectangle or a circle in two dimensions. Various techniques can then be used so that the computing time can be reduced significantly compared to general geometries. Direct solution methods based on Gauss elimination for a discrete system with N unknowns normally requires $\mathcal{O}(N^3)$ arithmetic operations. By using the fast Fourier transform for a problem defined on a rectangle or its generalizations in higher dimensions, this operation count can be reduced to $\mathcal{O}(N \log N)$. Similar reductions can be obtained for problems defined on a disk and its higher dimension generalizations. Therefore, for the model problem shown in Fig. 5.21, each step in the Schwarz iteration is computed very fast. However, note that the grid points in the subdomains in general don't coincide in the common area, and in particular not at the boundaries Γ_1' and Γ_1'. This means that an interpolation must be done when transferring from one subdomain to another.

The subdomain overlap introduces extra work in the sense that certain areas are treated several times as part of different subdomains. The alternative is non-overlapping domains, which early became the most common technique. One way of considering these methods is to start with discretizing the full problem without substructuring, and then partition the unknowns u_j into subsets that correspond to geometrical subdomains. Since the variables u_j are coupled to each other, the system must be modified so that each subsystem can be solved independently. For simplicity, we again consider the case with two subdomains. The partition is defined by the restriction matrices R_1, R_2:

$$\mathbf{u}_1 = R_1 \mathbf{u} = \begin{bmatrix} I & 0 \end{bmatrix} \mathbf{u}, \qquad \mathbf{u}_2 = R_2 \mathbf{u} = \begin{bmatrix} 0 & I \end{bmatrix} \mathbf{u}.$$

If the original system is

$$A\mathbf{u} = \mathbf{F}, \tag{5.30}$$

we define the submatrices

$$A_{11} = R_1 A R_1^T, \qquad A_{22} = R_2 A R_2^T,$$

and consider decoupled systems of the type

$$A_{11}\mathbf{u_1} = \mathbf{b}_1, \qquad A_{22}\mathbf{u_2} = \mathbf{b}_2$$

for any right-hand side $\mathbf{b}$. The solution to both systems combined can be written in the form

$$\mathbf{u} = (R_1^T A_{11}^{-1} R_1 + R_2^T A_{22}^{-1} R_2)\mathbf{b}.$$

The matrix on the right can now be used as a preconditioner for the original system (5.30):

$$(R_1^T A_{11}^{-1} R_1 + R_2^T A_{22}^{-1} R_2)A\mathbf{u} = (R_1^T A_{11}^{-1} R_1 + R_2^T A_{22}^{-1} R_2)\mathbf{F}.$$

This system has two independent systems that can be solved in parallel. Any iterative method with a structure that does not require the solution of any fully coupled system can now be applied. For example, the simple iteration

$$\mathbf{u}^{n+1} = \mathbf{u}^n + (R_1^T A_{11}^{-1} R_1 + R_2^T A_{22}^{-1} R_2)(\mathbf{F} - A\mathbf{u}^n)$$

can be used with each processor kept busy all the time, working only on the short vectors (convergence must of course be assured).

Another principle is used in the *Schur complement method* named after the Russian mathematician Issai Schur (1875–1941), another Jewish mathematician who encountered a lot of difficulties when working in Germany. His method is based on a grid which is partitioned into inner points and boundary points for each subdomain. In the simplest case discussed above with zero Dirichlet data on the outer boundary and two subdomains, the system has the form

$$\begin{bmatrix} A_{11} & 0 & A_{1\Gamma} \\ 0 & A_{22} & A_{2\Gamma} \\ A_{\Gamma 1} & A_{\Gamma 2} & A_{\Gamma\Gamma} \end{bmatrix} \begin{bmatrix} \mathbf{u}_1 \\ \mathbf{u}_2 \\ \mathbf{u}_\Gamma \end{bmatrix} = \begin{bmatrix} \mathbf{F}_1 \\ \mathbf{F}_2 \\ \mathbf{F}_\Gamma \end{bmatrix},$$

where Γ denotes the inner shared boundary. Unlike the Schwarz method, there is no overlap between the two subdomains here. The solution at the inner boundary satisfies the equation

$$Cu_\Gamma = \mathbf{F}_\Gamma - A_{\Gamma 1}A_{11}^{-1}\mathbf{F}_1 - A_{\Gamma 2}A_{22}^{-1}\mathbf{F}_2,$$

where C is the *Schur complement matrix*

$$C = A_{\Gamma\Gamma} - A_{\Gamma 1}A_{11}^{-1}A_{1\Gamma} - A_{\Gamma 2}A_{22}^{-1}A_{2\Gamma}.$$

The computations of $A_{11}^{-1}\mathbf{F}_1$ and $A_{22}^{-1}\mathbf{F}_2$ are carried out by solving two independent systems, and the point here is that these computations can be done in parallel.

It is obvious how to generalize the method to an arbitrary number of subdomains. However, for finite difference methods it is not trivial how to generalize to higher order accuracy, since in general such methods involve more than one single inner shared boundary. A possibility to get around this difficulty is to use deferred correction methods for raising the accuracy as suggested, see [81]. Deferred correction methods can be traced back to the article [55] by L. Fox and E.T. Goodwin published in 1949. Later Victor Pereyra picked up the idea and generalized it in a series of papers beginning with [134]. We indicate the basic principle by applying it to the trivial one-dimensional differential equation

$$\frac{\partial^2 u}{\partial x^2} = F(x)$$

complemented by boundary conditions. The standard second order difference approximation

$$Q_2 u_j = \frac{1}{h^2}(u_{j+1} - 2u_j + u_{j-1})$$

has a truncation error $(h^2/12)\partial^4 u/\partial x^4 + \mathcal{O}(h^4)$. The straightforward technique is to subtract a discrete version of this error giving the fourth order accurate approximation

$$Q_4 u_j = Q_2 u_j - \frac{h^2}{12}Q_2^2 u_j.$$

However, the deferred correction method is a two stage procedure where the first stage solution is used for the correction term:

$$Q_2 u_j^{(2)} = F_j,$$

$$Q_2 u_j^{(4)} = F_j + \frac{h^2}{12}Q_2^2 u_j^{(2)}.$$

These equations are applied at the inner points of each subdomain, and the left-hand side includes only the point at the interface at each side. On the right-hand side of the second equation, the operator Q_2^2 needs an extra point on the other side of the interface. This is a less severe complication since we are dealing with the already computed values $u_j^{(2)}$. However, an even more convenient procedure is to define these values by high order extrapolation from the inside.

Research in domain decomposition methods got an enormous boost in the late 1980s. The first "International Symposium on Domain Decomposition Methods" took place in 1987, and has been followed by 21 more, all of which attract top researchers. Maksymilian Dryja was one of the pioneers when these methods resurfaced. He wrote the first paper [45] on this topic in 1981, and he and Olof Widlund became leading researchers in this field with [46] as one of their first joint publications. Petter Bjørstad, Tony Chan, Roland Glowinski, Pierre-Louis Lions, and Barry Smith are others with important contributions.

Many survey articles and books on domain decomposition have been written by now, one of them is [153].

Olof Widlund
(Oberwolfach Photo
Collection (MFD); photo by
Renate Schmid (MFO).
https://owpdb.mfo.de/)

5.13   Wave Propagation and Open Boundaries

Different types of wave propagation are common in nature, and there is an ever increasing need for numerical simulations associated with acoustics, electromagnetics and geophysics. The mathematical models are well known as systems of partial differential equations, but except for very simple cases, it is not possible to find any explicit analytic form for solutions that can be used for computation. Numerical methods are also well known, but there is often a remaining difficulty. We take acoustics as a typical example.

Assume that there is a sound source, for example a windmill, and one wants to compute the noise level at a certain village located at some distance from the windmill. The sound propagates in an unbounded three-dimensional domain, but a direct simulation based on a numerical approximation of the wave equation must be limited to a finite domain with an artificial nonphysical boundary. The question then arises how the boundary conditions should be constructed there. At the ground surface they can be defined, at least in principle, by using physical laws for reflection depending on the properties of the surface. However, what should we do at an imagined surface in the air that is the boundary of the computational domain? The real sound waves don't notice any obstacle in the air; they just continue propagating without reflections.

Björn Engquist   (Courtesy
Björn Engquist)

In the 1970s Heinz Kreiss had a bright student Björn Engquist (1945–) at Uppsala University. After finishing his Ph.D. thesis he moved to Stanford University, where he worked for Jon Claerbout on wave propagation problems in geophysics. Soon he encountered the problem described above, but then for seismic waves propagating through the earth. At Stanford he met Andrew Majda (1949–), and they started working together on the boundary condition problem. This resulted in a general method for constructing what is usually called *absorbing boundary conditions*, see [49] published in 1977. We shall describe the method for the simplest possible problem in two space dimensions with plane waves propagating with speed one, which are governed by the wave equation

$$\frac{\partial^2 u}{\partial t^2} = \frac{\partial^2 u}{\partial x^2} + \frac{\partial^2 u}{\partial y^2},$$

$$u(x, y, 0) = f(x, y).$$

If the problem is defined in the whole x, y-space, we now introduce a straight line $x = x_0$ as an artificial boundary and compute the solution in the domain $\{(x, y)/x \geq x_0\}$. A plane wave has the form

$$u(x, y, t) = e^{i(\xi x + \eta y + \omega t)},$$

where ξ and η are the wave numbers in space, and ω is the frequency in time. The wave equation provides the dispersion relation

$$\omega^2 = \xi^2 + \eta^2, \tag{5.31}$$

which can be seen as the Fourier transformed version. When solving for ξ, we get

$$\xi = \pm\omega\sqrt{1 - \eta^2/\omega^2},$$

Andrew Majda
(Photographer: George M.
Bergman. Oberwolfach Photo
Collection. Illustration from
Wikimedia*)

which allows for all wave directions. We consider now waves that are going from right to left through the boundary. This is obtained with the plus sign and $\xi/\omega > 0$, which is the exact condition for left-going waves. If the boundary is present for computational reasons only, and we know that there are no right-going waves, this condition should be enforced in order to avoid any nonphysical reflections.

The problem is that the condition

$$\xi = \omega\sqrt{1 - \eta^2/\omega^2}, \qquad |\eta/\omega| \leq 1 \tag{5.32}$$

is given in Fourier space, and we want a condition in physical space where the computation is done. The dispersion relation (5.31) corresponds to the relation

$$\frac{\partial^2}{\partial t^2} = \frac{\partial^2}{\partial x^2} + \frac{\partial^2}{\partial y^2},$$

but the condition (5.32) does not correspond to any relation between differential operators as we know them. However, the right-hand side is a pseudo-differential operator, and since it is an algebraic expression, we can approximate it such that the condition corresponds to a relation between genuine differential operators.

The assumption is that $|\eta/\omega|$ is small, i.e., the wave has little variation in the y-direction. In the special case that $\omega = 1$, one can interpret η as $\eta = \sin\theta$, where

θ is the angle of incidence of the wave. Hence approximations of ξ in (5.32) can be constructed with different orders of accuracy in $|\eta/\omega|$. With an increasing order of accuracy, the reflections back into the domain become smaller.

The first approximation

$$\sqrt{1 - \eta^2/\omega^2} = 1 + \mathcal{O}(\eta^2/\omega^2)$$

implies

$$\xi = \omega,$$

which after multiplication by i leads to the boundary condition

$$\frac{\partial u}{\partial t} - \frac{\partial u}{\partial x} = 0, \qquad x = x_0.$$

The next approximation

$$\sqrt{1 - \eta^2/\omega^2} = 1 - \frac{\eta^2}{2\omega^2} + \mathcal{O}\left(\frac{\eta^4}{\omega^4}\right)$$

leads to the boundary condition

$$\frac{\partial^2 u}{\partial x \partial t} - \frac{\partial^2 u}{\partial t^2} + \frac{1}{2}\frac{\partial^2 u}{\partial y^2} = 0, \qquad x = x_0.$$

The authors then continue with even higher order approximations leading to weaker reflection waves.

By introducing boundaries, the original initial value problem is transformed into an initial boundary value problem, and the well-posedness of this new problem must be verified. The two conditions above do not introduce any trouble with well-posedness, but for higher order, one has to choose the type of approximation carefully to avoid ill-posedness.

The absorbing boundary conditions are derived for the differential equation, but for computation, the whole problem must be discretized, and stability must be investigated. For the standard most compact difference approximation of the wave equation with the two boundary conditions above, there are no stability problems with either one.

The technique for deriving absorbing boundary conditions can be generalized to hyperbolic systems of PDE, and this is done in the original paper [49].

As mentioned above, there are some stability difficulties for high order approximations. Despite this fact, we have presented the derivation in some detail, the reason being that the authors introduced a very interesting new technique. In the past pseudo-differential operators had been derived and analyzed by researchers in the pure mathematics community. Engquist and Majda showed that they can be used also for applications in computational mathematics.

The Engquist–Majda technique was a new way of handling the problem arising with the introduction of artificial boundaries. However, several attempts have been made to develop other techniques. The most important and very different technique was invented by Jean-Pierre Berenger and presented in [8] in 1994. It was developed for Maxwell's equations for electromagnetic wave propagation, and we give a brief summary of the principle.

The equations are written as a first order system, and in two space dimensions they are

$$\epsilon_0 \frac{\partial E_x}{\partial t} + \sigma E_x = \frac{\partial H_z}{\partial y},$$

$$\epsilon_0 \frac{\partial E_y}{\partial t} + \sigma E_y = \frac{\partial H_z}{\partial x},$$

$$\mu_0 \frac{\partial H_z}{\partial t} + \sigma^* H_z = \frac{\partial E_x}{\partial y} - \frac{\partial E_y}{\partial x}$$

for the electric field E_x, E_y and the magnetic field H_z, in which the parameters σ and σ^* are the electric and magnetic conductivity respectively. For vacuum $\sigma = \sigma^* = 0$, in this case the waves are moving with the speed of light $c = 1/\sqrt{\epsilon_0\mu_0}$.

The basic idea is to add an extra layer outside the primary domain of interest and then choose the parameters σ, σ^* so that the waves pass into the layer without reflection, and then become strongly damped before they reach the outer boundary of the new layer. Figure 5.22 shows the setup. In the original version, the magnetic

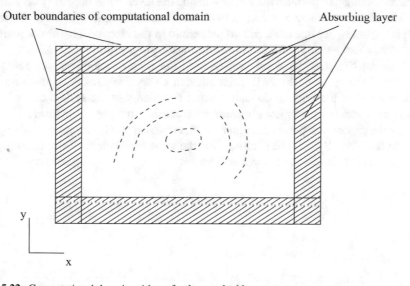

Fig. 5.22 Computational domain with perfectly matched layers

field is split into two components H_{zx}, H_{zy} satisfying the extended system

$$\epsilon_0 \frac{\partial E_x}{\partial t} + \sigma_y E_x = \frac{\partial (H_{zx} + H_{zy})}{\partial y},$$

$$\epsilon_0 \frac{\partial E_y}{\partial t} + \sigma_x E_y = \frac{\partial (H_{zx} + H_{zy})}{\partial x},$$

$$\mu_0 \frac{\partial H_{zx}}{\partial t} + \sigma_x^* H_{zx} = -\frac{\partial E_y}{\partial x},$$

$$\mu_0 \frac{\partial H_{zy}}{\partial t} + \sigma_x^* H_{zy} = \frac{\partial E_x}{\partial y}.$$

If the conditions

$$\sigma_x / \epsilon_0 = \sigma_x^* / \mu_0,$$

$$\sigma_y / \epsilon_0 = \sigma_y^* / \mu_0$$

are satisfied, then σ_x, σ_y can be chosen such that the magnitude of the waves are affected by an exponential damping factor of the type $\sim e^{\alpha y}$, $\alpha < 0$ in the upper layer, and similarly in the other three layers. (The construction requires a special choice of parameters in the four corner subdomains, such that the two media interfaces at each corner do not cause any extra trouble.)

In general, the solution is not quite zero at the outer boundary, and boundary conditions are required here. These are often chosen so that they correspond to perfect conduction, i.e., the proper component of the electric field is set to zero. This allows for a reflected wave, but one can show that also this wave will be damped. The remaining part pollutes the solution inside the layer, but is hopefully very small.

The Beringer method, often called the PML-technique (Perfectly Matched Layer), quickly became the standard procedure in the electromagnetic community, and it was also quickly generalized to other wave propagation problems.

Without damping the wave propagation, the time development preserves the energy of the system. The PML-principle can be seen as modifying this property which holds in the original domain, so that the energy of the system decreases in the extra layer. In its simplest illustration we consider a wave of the form $e^{i(\xi x - \omega t)}$, where the exponent is purely imaginary. If an extra real part is introduced, a damping can be achieved. By analytic continuation, the wave can be defined also for complex $\tilde{x} = x + i g(x)$, where $g(x)$ is real and we get

$$\partial \tilde{x} = (1 + i \frac{dg}{dx}) \partial x.$$

Considering the real part of $\tilde{x}$, we choose $df/dx = \sigma(x)/\omega$ and make the substitution,

$$\frac{\partial}{\partial x} \rightarrow \frac{1}{1 + i\sigma(x)/\omega} \frac{\partial}{\partial x}.$$

The space part $e^{i\xi x}$ of the wave now becomes

$$e^{i\xi x} e^{-\frac{\xi}{\omega} \int^x \sigma(x')dx'},$$

showing that for $\sigma = 0$, there is no change, but for $\sigma > 0$, there is a damping in the positive x-direction.

This procedure can now be modified so that it applies in the negative x-direction as well as in the positive and negative y-directions. The final result is a procedure that leads to the same type of extra layers, analogous to the PML-principle. A good and more detailed description of generating damping via a coordinate transformation is found in *Notes on Perfectly Matched Layers*, published by S.G. Johnson as course material for an MIT course and available on the web.

As always, the problem must be discretized for computational purposes. And as always, the discrete procedure must be analyzed with respect to stability.

Finally a remark on the labelling of the two techniques. Engquist and Majda already called their original method *absorbing boundary conditions* in the title of the article. But the label *nonreflecting boundary conditions* is often used, and it is actually a better description. The idea is to let the waves pass through the artificial boundaries without generating any reflected waves. The PML-technique is also constructed so that no reflections occur at the inner boundary of the extra outer domain, but outside of this boundary, the waves are *absorbed* in the true sense by the artificial medium.

5.14 Stochastic Procedures for Deterministic Problems

5.14.1 Monte Carlo Methods

The label Monte Carlo methods refers to the gambling casino in Monaco and indicates that we are dealing with random numbers and probability. How can this have anything to do with the precise science of computation?

We have again a class of methods whose source goes back to the Manhattan Project and the following activities at Los Alamos Scientific Laboratory. Stanislaw Ulam (1909–1984) was a Polish Jewish mathematician who emigrated to United States 10 days before the outbreak of World War II. He held positions at different universities, and spent long periods at Los Alamos working with John von Neumann among others.

In 1946 Ulam became seriously ill, but recovered after a resting period in his home. To get time to pass faster, he used to play solitaire with a deck of cards. Given the mathematician that he was, he became interested in calculating the probability that the patience game would come out to be winnable. This problem led to a very complicated analysis of all possible combinations, and he got the idea that a much simpler way would be to simulate the patience procedure repeatedly by choosing to shuffle and distribute cards randomly. He then thought of the possibility of applying the same principle to solve the problem of neutron diffusion which had high priority at Los Alamos in constructing sufficiently safe shielding. He told von Neumann about the idea, and this was the start of their work on the Monte Carlo method. It was classified work and nothing was published until 1949, when Ulam and Nicholas Metropolis (1915–1999) wrote the first paper [119] on the method. It is a rather unusual mathematical paper, since it contains very few mathematical expressions, but rather a wordy explanation of the principles for the method, much as in the old days.

The outcome of patience and the process of neutron scattering are both examples of problems with a basic discrete description, in which each event has an associated probability distribution. However, Ulam and his coworkers saw the connection with corresponding macroscopic equations like the diffusion differential equation in the case of neutron scattering. As a consequence, one can start by considering a given macroscopic mathematical model, and then discretize it for computational purposes. This is the principle for almost all numerical methods, but the difference here is that the discretization is not only nonuniform, but also determined by random numbers. Let us demonstrate it for computing integrals.

Stanislaw Ulam   (©Emilio
Segre Visual
Archives/American Institute
Of Physics/Science Photo
Library; 11831352)

In Sect. 3.3 we described classical methods for computing integrals by using interpolation on regular grids. In one space dimension and the integral

$$I = \int_0^1 f(x)\,dx,$$

these methods have the form

$$I_\Delta = \sum_{j=1}^{n} \alpha_j f(x_j)\, \Delta x,$$

where α_j are certain constants and $n\Delta x = 1$. The error has the form

$$|I_\Delta - I| \le C\Delta x^p,$$

where C is a constant and p depends on the degree of interpolation. For example, Simpson's rule has an order of accuracy $p = 4$.

Consider next an integral in two space dimensions

$$I = \int_0^1 \int_0^1 f(x, y)\, dx dy$$

and its approximation

$$I_\Delta = \sum_{j=1}^{n_x} \sum_{k=1}^{n_y} \alpha_{jk} f(x_j, y_k)\, \Delta x \Delta y,$$

where $n_x \Delta x = n_y \Delta y = 1$. The error now has the bound

$$|I_\Delta - I| \le C(\Delta x^p + \Delta y^p) = C\left(\frac{1}{n_x^p} + \frac{1}{n_y^p}\right).$$

With the same total number n of f-evaluations as in the one-dimensional case, we have $n_x = n_y = \sqrt{N}$, giving the error estimate

$$|I_\Delta - I| \le \frac{C}{N^{p/2}}.$$

In d space dimensions, this generalizes to

$$|I_\Delta - I| \le \frac{C}{N^{p/d}},$$

which results in poor accuracy if d is large. Another way of seeing this is by first considering the 1-D case and choose N such that reasonable accuracy is obtained. In d dimensions and the same accuracy requirements, we then need n^d evaluation points, which for increasing d quickly becomes out of reach even on modern computers.

The Monte Carlo method does not use any predetermined distribution of points. Instead each evaluation point is picked at random in the d-dimensional hypercube

$[0, 1] \times [0, 1] \times \ldots [0, 1]$. With $\mathbf{x}_j$ denoting a random vector with d elements representing the coordinates in the hypercube, the integral is

$$I = \int_0^1 \int_0^1 \cdots \int_0^1 f(\mathbf{x}) \, dx_1 dx_2 \ldots dx_d$$

with the numerical approximation

$$I_{MC} = \frac{1}{n} \sum_{j=1}^n f(\mathbf{x}_j).$$

It may look like a crude approximation. Not only do we use random numbers, but there is also a complete lack of interpolation formulas for the function $f(\mathbf{x})$. So why is this a useful method?

The reason can be traced back to the central limit theorem in probability theory, which is the basis for the error estimate. Let $\{x_j\}$ be a set of random numbers, each with average $E(x_j) = \mu$ and variance σ^2. The theorem says that

$$\frac{1}{n} \sum_{j=1}^n x_j - \mu \to^d \frac{N(0, \sigma^2)}{\sqrt{n}},$$

where $N(0, \sigma^2)$ denotes the normal distribution, and the limit process is in the distribution sense. The Monte Carlo method can then be proven to have similar convergence property, i.e.,

$$|I_{MC} - I| \le \frac{C}{\sqrt{n}},$$

which is independent of d. This is slow convergence compared to many other classes of problems, but as we saw above, it is in practice the only possible method for large d.

We shall now discuss the solution of PDE, where parabolic equations are most suitable for the Monte Carlo method, in particular the heat equation. The procedure is based on the *random walk*; here each step is controlled by a random number. This method has an interesting history, and can be traced back to the English statistician Karl Pearson (1857–1936). In a letter to *Nature* in 1905, he posed the following problem: A drunk person starts at a certain point and takes a step of length d in a random direction, then he turns around a random angle and takes another step of length d. Then he proceeds in the same manner and the question is where he ends up after n steps with the answer given as a probability distribution of the final position as a function of the distance r from the original point. The English physicist Lord Raylegh (1842–1919) answered quickly in another letter to *Nature* that he had already solved essentially the same problem in 1880. One result of the analysis is that the drunkard usually ends up quite close to where he started.

Independently, about the same time, the French mathematician Louis Bachelier (1870–1946) used random walk methods for problems in financial mathematics, and he also found the connection between these methods and the heat equation. This was the basis for the much later development of Monte Carlo methods for more general partial differential equations.

We illustrate the method by the simplest form of the heat equation,

$$\frac{\partial u}{\partial t} = \frac{\partial^2 u}{\partial x^2}, \tag{5.33}$$

and discretize it on a uniform grid with stepsize Δt in time and Δx in space. We think of $u(x, t)$ as the probability that a particle is located at x at time t, and assume first that at $t = 0$, it is 1 at $x = x_0$ and zero everywhere else. This initial condition can be written as

$$u(x, 0) = \delta(x - x_0),$$

where $\delta(x)$ is the Kronecker delta function which is one at $x = 0$ and zero otherwise. The idea is now to proceed with a random walk procedure with stepsize Δx, where a random number is drawn with equal probability of moving left or right. Then it is easily shown by the Taylor expansion that with decreasing stepsizes and

$$\Delta x^2 / \Delta t \rightarrow 1,$$

the procedure can be seen as an approximation of the differential equation (5.33) at $x = x_0$.

Denote the solution of (5.33) with the initial condition $u(x, 0) = \delta(x)$ by $K(x, t)$. Then the solution with the initial condition

$$u(x, 0) = u_1 \delta(x - x_1) + u_2 \delta(x - x_2) + \ldots + u_n \delta(x - x_n)$$

is

$$u(x, t) = u_1 K(x - x_1, t) + u_2 K(x - x_2, t) + \ldots + u_n K(x - x_n, t).$$

This is an approximation of the true solution in the form

$$u(x, t) = \int K(x - \xi) u_0(\xi) \, d\xi.$$

The method is therefore to start at the point x_1, and then make N random walks with each trajectory consisting of M timesteps. The x-axis is divided into L subintervals I_k that are usually larger than the stepsize h, and the number of trajectories $v_{k,1}$

ending up in each subinterval I_k is registered. The same procedure is then repeated for each point $x_2, x_3, \ldots, x_n$ giving $v_{2,k}, v_{3,k}, \ldots, v_{n,k}$ as the number of endpoints for each subinterval I_k. The final solution at $t = t_M$ is then the weighted sum

$$u(k, t_M) = \frac{1}{N} \sum_{j=1}^{n} u_j v_{j,k}, \qquad k = 1, 2, \ldots, L,$$

representing the solution in the interval I_k.

The problem with the initial function

$$u(x, 0) = \begin{cases} 0, & |x| > 1, \\ \sin\big((x + 1)\pi/2\big), & |x| \leq 1 \end{cases}$$

is solved with $h = 0.02$ and length 0.1 of each subinterval. The solution at $t = 0.5$ is shown in Fig. 5.23 for three different cases $N = 500, 1000, 2000$ random walks respectively.

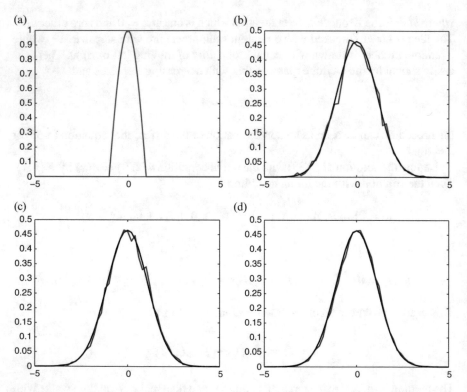

Fig. 5.23 Monte Carlo solution (red), exact solution (black) at $t = 0.5$. (Note the different scale for the initial function.) (**a**) Initial function at $t = 0$. (**b**) N = 500. (**c**) N = 1000. (**d**) N = 2000

The figure shows that quite a large number of random walks is necessary for acceptable accuracy, and a standard finite difference method is much faster. However, the method generalizes in a straightforward manner to d space dimensions, and there it may be competitive under certain conditions. A point $\mathbf{x}$ is defined by the Cartesian coordinates as

$$\mathbf{x} = \begin{bmatrix} x_1 \\ x_2 \\ \vdots \\ x_d \end{bmatrix}.$$

The d-dimensional equation becomes

$$\frac{\partial u}{\partial t} = \alpha \left(\frac{\partial^2 u}{\partial x_1^2} + \frac{\partial^2 u}{\partial x_2^2} + \ldots + \frac{\partial^2 u}{\partial x_d^2} \right),$$

and the random walk takes place in the d-dimensional space with the stepsize Δx in each direction, where

$$\Delta x^2 / \Delta t = 2d\alpha.$$

The initial condition is defined by n points as

$$u(\mathbf{x}, 0) = u_1 \delta(\mathbf{x} - \mathbf{x}_1) + u_2 \delta(\mathbf{x} - \mathbf{x}_2) +, \ldots, u_n \delta(\mathbf{x} - \mathbf{x}_n).$$

At a given point $\mathbf{x}_0$, there are $2d$ directions to choose from, and they are given equal probability $1/(2d)$ for each step. The algorithm follows now the same pattern as in the 1D-case, with the final solution $u_k(t_M)$ computed as a weighted sum over n points $\mathbf{x}_j$.

The method can be generalized to more general PDE of a parabolic character and even to other types of PDE. Furthermore, the method must be modified when boundaries are present, where one procedure is to simply cancel every trajectory that is leaving the domain.

Also for PDE solvers the Monte Carlo method has poor accuracy but, as for integrals, it becomes more attractive for problems in many dimensions.

Monte Carlo methods require access to random numbers, usually with a statistical uniform distribution. A long time ago, lists of random numbers were available, but they were of course of no practical use on a computer. Again, von Neumann came up with a new method, which generated *pseudo-random numbers*. They are not truly random; indeed they are completely deterministic once a set of parameters has been selected. The point is that they have properties that are very close to truly random numbers, they can be generated fast, and they serve the purpose required by the Monte Carlo method. Things can go wrong, but von Neumann argued that if that happens, it is obvious to the user and can then be corrected. In fact, much effort has been devoted to the problem of generating pseudo-random numbers over the years, and the standard generators are very reliable today.

5.14.2   An Example from Financial Mathematics

Monte Carlo methods are frequently used in a wide variety of application areas. One of the most important is financial mathematics, in which stochastics and probabilities come in as natural ingredients in mathematical models. As an example, we consider option pricing. If S_t denotes the price process of an asset, the governing model is the stochastic differential equation

$$dS_t = \mu S_t dt + \sigma S_t dw_t, \tag{5.34}$$

where μ is the drift and σ is a measure of the variation of the stock price called volatility. The last term represents the stochastic part, where w_t denotes Brownian motion. This concept is named after Robert Brown (1773–1858), who observed the random motion of particles in a fluid. The mathematical definition has later been applied to many different areas, and the equation above is one of them.

An option issued on the underlying asset (5.34) with pay-off function $\Phi(S_T)$ at the time of maturity T has the value

$$V = e^{-rT} \int_{-\infty}^{\infty} \Phi(S_T) dS_T,$$

today, where e^{-rT} represents discounting from the time of maturity to present time with the risk-free interest rate r. The distribution of S_T is obtained by discretizing (5.34) on a grid in time and using pseudo-random numbers for the last part. Because of the stochastic ingredient, the procedure is called the Monte Carlo method. The expected value V is computed as the average over N simulated trajectories of S as

$$V = e^{-rT} \frac{1}{N} \sum_{j=1}^{N} \Phi(S_T(j)).$$

It seems that there is no alternative to the stochastic approach to this problem. However, in 1973 Fischer Black (1938–1995) and Myron Scholes (1941–) derived an equivalent PDE for the price $V(S, t)$ of a derivative as a function of a stock price and time, see [10]. The same year Robert Merton (1944–) published a paper referring to this work (writing "forthcoming" in the reference list), in which he works out weaker conditions on the underlying necessary assumptions. Scholes and Merton were awarded the Nobel Memorial Prize in Economic Sciences in 1997 for their work; Black died in 1995 and the prize was not awarded posthumously. The model came to be called the Black–Scholes equation

$$\frac{\partial V}{\partial t} + \frac{\sigma^2 S^2}{2}\frac{\partial^2 V}{\partial S^2} + rS\frac{\partial V}{\partial S} - rV = 0,$$

$$V(S, T) = \phi(S),$$

where r is the risk-free interest rate. When solving backwards in time, this is a nonlinear parabolic PDE, and offers the possibility to use efficient numerical PDE solvers for a solution. Actually, it is often said that this is the most frequently solved differential equation in all categories.

Does the existence of this equation mean that the Monte Carlo method as described above is out of the question nowadays? No, it doesn't. The reason is that there is much interest in portfolios containing several different stocks, which leads to PDE in several space dimensions. And there we go again. "The curse of dimensionality" hits again, i.e., for a large number of stocks, PDE-methods on traditional grids in each dimension cannot be used. The Monte Carlo method is one possible alternative choice, even if the convergence rate is again only of the order $1/\sqrt{n}$.

We also note that the random walk version of the Monte Carlo method as described in the previous section can be applied to the Black–Scholes equation, and the parabolic character makes it particularly suitable for this method.

Finally, we note that the methods based on radial basis functions described in Sect. 5.5.3 are well suited for problems in many space dimensions, and the Black–Scholes equations are one of the main targets for these methods.

A survey and comparison of all currently used methods is found in [159].

5.15   Level Set Methods

Computing PDE solutions that have discontinuities is very difficult. We have discussed this problem earlier in this book, in particular the problem with shocks in fluids, see Sect. 4.3.5. Standard finite difference methods and finite element methods generate oscillations around the discontinuities, and we have discussed how this can be overcome either by avoiding operators that are applied across the discontinuity, or by using methods with inherent damping. However, in 1988 Stan Osher and James Sethian introduced the *level set method*, which is based on a different principle, see [128]. It can be used as the basis for solving many different kinds of problems, and for simplicity, we choose here the image processing problem for illustration.

In two space dimensions we assume that there is an interface Γ which is the boundary of the open domain Ω. There is also a velocity field represented by the vector $\mathbf{v} = [v_x, v_y]^T$ governing the movement of Γ, which is associated with time t. The velocity depends on (x, y, t) and possibly also on the shape of the interface. The main idea is to define a smooth function $\phi(x, y, t)$ with the following properties:

$$\phi(x, y, t) > 0, \qquad (x, y) \in \Omega,$$

$$\phi(x, y, t) < 0, \qquad (x, y) \notin \bar{\Omega},$$

$$\phi(x, y, t) = 0, \qquad (x, y) \in \Gamma.$$

The connection between ϕ and Γ is illustrated in Fig. 5.24, where the red surface represents a changing function ϕ at three different stages, the middle one showing the point in time where Ω breaks into two pieces.

With the notation $\nabla\phi = [\partial\phi/\partial x, \partial\phi/\partial y]^T$, it can be shown that the motion is governed by the equation

$$\frac{\partial\phi}{\partial t} + \mathbf{v}^T \nabla\phi = 0,$$

where $\mathbf{v}$ is the velocity at the interface, otherwise arbitrary. If $\| \cdot \|$ denotes the usual Euclidean norm of a vector, the normal velocity component is

$$\mathbf{v}_N = \mathbf{v}^T \frac{\nabla\phi}{\|\nabla\phi\|},$$

giving the basic equation

$$\frac{\partial\phi}{\partial t} + \mathbf{v}_N \|\nabla\phi\| = 0. \qquad (5.35)$$

Fig. 5.24 Connection between ϕ and Γ. (Creator: Oleg Alexandrov at English Wikipedia. Illustration from Wikimedia*)

The numerical solution of this equation is the kernel of the level set method. With the function ϕ known at all gridpoints in the (x, y)-plane at a given time t, the interface is obtained by finding out where ϕ is zero. Since ϕ is a smooth function, this is a convenient procedure compared to the alternative in which each point of the interface is advanced using the velocity field in all situations including the case where Γ breaks into two pieces. This is not to say that it is easy to solve (5.35) in the nonlinear case where $\mathbf{v}$ depends on the properties of Γ. However, methods for this type of problems exist, for example high order ENO (Essentially Non-Oscillatory) methods, see [129].

Image processing is a very common working tool in various types of high technology applications, many of them being used in daily life, and the processing techniques are nowadays very often based on the level set method. We refer to the article [154] for an excellent review of the topic.

Chapter 6
Impact of Numerical Analysis and Scientific Computing

As we have seen in this book, computational methods have a long history, starting with the Babylonians. Some analytical techniques and numerical algorithms have had a particularly strong impact, and in this final section we shall point out some of those that are still used with significant impact on computational science.

We note here that handling large databases in an optimal way is very common nowadays. Perhaps the most well known among these algorithms are those extracting relevant(?) information from an enormous database, often with probability distributions involved. This is of interest for just about everyone these days and has a strong impact in daily life, but these methods are not discussed here.

6.1 Numerical Methods

Euler was a mathematical giant, and he was probably the first to use a **finite difference method** for solving a differential equation. For systems of ordinary differential equations, a long series of variations on the basic principle was later developed with Runge–Kutta methods as one of the most important, in particular the explicit versions. Curtiss and Hirschfelder constructed the backwards differential formulas (BDF), and these are today standard methods for stiff systems of ODE.

The first person to use finite difference methods for PDE was Richardson, even if he failed to achieve any accurate results due to the fact that he didn't know about stability for time-dependent problems. After the war, these methods dominated this field, both for initial and boundary value PDE problems.

In the 1950s a new approach for PDE showed up, namely the **finite element method**. Courant had already used it for a simple problem, but the real start for realistic problems came in the aircraft industry with Jon Turner as the inventor. The more general versions developed later are based on the Ritz–Galerkin formulation which had been introduced in the beginning of the twentieth century. The method

© Springer International Publishing AG, part of Springer Nature 2018   233
B. Gustafsson, *Scientific Computing*, Texts in Computational Science
and Engineering 17, https://doi.org/10.1007/978-3-319-69847-2_6

had a strong period of development, and is today the standard method for all sorts of PDE problems. One reason for this is its flexibility when it comes to approximating irregular domains.

With the introduction of parallel computers, and later massively parallel computers came the need for numerical methods which could be partitioned into different parts that would keep each processor busy simultaneously. Schwarz was early out with his alternating method for an elliptic PDE with two subdomains, but in 1988 Lions presented the general **domain decomposition method** which was later modified in various ways.

With the PDE solvers came a need for a quick solution for large linear systems of algebraic equations. Gauss was the first to use iterative methods, and the original **Gauss–Seidel method** is still used for some applications. However, there are two other methods that dominate today. One is the **multigrid method** due to Fedorenko, but further developed by Brandt and Hackbusch. The other class involves **Krylov space methods** in various versions, one of them called GMRES.

Nonlinear problems of algebraic equations are usually solved by the **Newton–Raphson method**. The original method was invented a long time ago by Newton, but more than 400 years later, it is still the standard technique for most applications. This is an illustration of the fact that the greatest mathematicians often are behind the most important numerical methods. The method can be applied to systems of algebraic equations as well, and since linearization is the basic principle, the methods for a solution to linear systems of algebraic equations described above are applied at each stage of the Newton iteration.

The **least squares method** is frequently used for representing functions having fewer parameters than the number of given data. It was invented by Gauss, and it is certainly the standard method in almost every kind of application area when it comes to representing measured data by continuous functions. It also has particular significance in statistical problems.

Many lists of the most important algorithms of all time have the **FFT** for computing the discrete Fourier transform at the top. For N given data points, it reduces the operation count from $\sim N^2$ for straightforward algorithms to $\sim N \log N$. The FFT has a large application area, not only in signal analysis, but also in various numerical methods as we have demonstrated above. It has been shown that Gauss had an algorithm essentially equivalent to the FFT, but it was not until Cooley–Tukey published their version in 1965 that it became used.

Problems containing scales of a very different order are very common. Important examples of efficient methods for such problems are the **Fast Multipole Method** developed by Rokhlin for computing integrals, and the **Heterogeneous Multiscale Method** developed by E and Engquist for differential equations. Both these method classes reduce problems of an almost impossible size to algorithms with a reasonable operation count.

Linear optimization has applications in many different areas. Kantorovich came with a general method in 1939, and it was the basis for the **Simplex method** developed by Dantzig during the war, but for secrecy reasons not published until 1947. It is still the standard method for all kinds of applications.

6.2 Analysis Techniques

Partial differential equations are frequent mathematical models in science and engineering, and except for very simple equations, they are impossible to solve by analytical means. Numerical methods obtained either by series expansion or by discretization lead to heavy computation, which made it almost impossible to get any reasonable results before electronic computers became available. Furthermore, the analysis of stability and convergence is quite complicated, and when listing significant achievements in numerical analysis, these kinds of problems play a central role.

Before the computer era, few attempts were made to solve time-dependent PDEs except for very short time-intervals, and the stability concept was essentially unknown. However, there was one very important prewar paper bringing up this issue, namely [30] published by Courant, Friedrichs and Lewy in 1928 which led to the **CFL-condition**, see 3.8.5. It was derived for the wave equation, but could easily be generalized to other hyperbolic equations. The practical impact of this condition is that the time-step must be chosen sufficiently small in order to include the domain of dependence for the differential equation in the domain of dependence for the numerical scheme. Almost a century later, the CFL-condition is still referred to when analyzing numerical methods for hyperbolic problems.

Another essential condition concerning stability is the **von Neumann condition**. Von Neumann may have been the most influential scientist ever when it comes to computational science. His method of Fourier transforming the PDE approximation in space and investigating the resulting difference scheme in the time-direction is still by far the most commonly used technique for stability analysis. There are two curious facts concerning this condition. The first is that von Neumann didn't publish it himself; it was done by Crank and Nicolson in their paper on a new difference method, although they were referring to Hartree who had communicated the technique to the authors, but in turn had "followed a suggestion by Prof. J. von Neumann". The other curious fact is that the condition is applied even if there is no theoretical basis for it. The underlying assumption is that the solution is periodic and expressed in terms of exponentials $e^{i\omega x}$, or alternatively, the solution is defined in the whole space without any boundaries. Such problems exist, but generally boundaries are present. Still, the von Neumann condition is the only one used for analysis in many cases, and even if it is not sufficient for stability, it seems to be a good indication for choosing the time-step for practical applications.

The **equivalence theorem** presented by Lax and Richtmyer in 1956 was a very deep result. However, it didn't lead to any new technique for analysis, but established the importance of stability from a theoretical point of view. The same thing can be said about the **matrix theorem** presented by Kreiss in 1962.

The theory for difference approximations of ODEs got a quick start after the war with further developments during the following decades. The theory developed for stiff ODE systems and the introduction of the concept **A-stability** by Dahlquist was of fundamental significance in many applications with the Dahlquist barrier as a well known concept for most practitioners.

The work of Kreiss based on the Laplace transform-technique gave sufficient conditions for the stability of difference schemes for initial-boundary value problems, today usually called the **GKS-condition**.

6.3   Examples from Computational Physics, Chemistry and Engineering

The use of numerical methods for problems in science and engineering is now so widespread and with so many applications that it is impossible to give any survey. We shall rather give a few examples to illustrate the impact on our understanding of basic properties in some selected areas. The two largest fields of science are physics and chemistry with engineering as another field with more immediate applications. We are choosing examples that are attracting a lot of attention and are having a strong impact on present day computational science.

6.3.1   Computational Physics

Physicists were early in their use of computational methods, the first being involved in the construction of the nuclear bomb in USA. Meteorology is an area where there has been a very intensive development for many years. The reason is obvious because of the fundamental interest in every country to have reliable weather predictions. In Sect. 3.8.4, we described the first prewar attempt by Richardson to predict the weather, using a difference method approximating a time-dependent PDE. Unfortunately, he didn't succeed, since he was using an unstable numerical scheme.

The existence of the electronic computer made a dramatic difference. The first computation was made by Jule Charney, Ragnar Fjotoft and John von Neumann in 1950; the publication [21], mentioned in Sect. 4.3.1, was remarkable for containing the first general description of Fourier analysis that ended up with the von Neumann condition. The mathematical model was simple in the beginning, but the speed using the ENIAC computer was high enough to keep up with weather progress just about at the same pace. Carl-Gustav Rossby at the Stockholm University used the barotropic vorticity equation and a better numerical method to achieve a simulation fast enough to make predicting weather of practical use. The calculation was made on the Swedish computer BESK, which was developed with Rossby as an active participant.

Making a big jump in time to present-day weather predictions, we conclude that calculations are now fast enough to produce quite accurate week-long forecasts. They are usually based on the primitive equations, which are a set of PDEs that couple such physical variables as flow velocity, pressure, density and temperature. Finite difference methods of fourth order accuracy are often used for numerical computation. One version is the global model defined over the whole globe, or alternatively the northern hemisphere. Interestingly, due to the periodic solutions in two out of three space dimensions, this is a case where a **pseudo-spectral method** can be used, see Sect. 5.8.1. Since such a method has very high accuracy, a relatively coarse grid can be used.

Weather prediction is a challenge that has been of interest to all mankind for centuries. Numerical methods are now quite advanced, and predictions are usually very accurate for a few days. However, detailed reliable predictions for more than 2 weeks cannot be provided for the foreseeable future. The reason is that the equations are nonlinear and chaotic behavior shows up. This is the famous butterfly effect which says that a small perturbation somewhere causes a much larger global effect after some time. This requires computation on such a small scale so that it is unrealistic even with the fastest and largest computers.

6.3.2 Computational Chemistry

In chemistry the basic question for a long time has been to understand various chemical reactions on an atomic scale, in particular how electrons are moving. The Schrödinger equation describes this mechanism and leads to quantum chemistry. For computation, an established model for atomic orbitals had been represented as a set of functions called the *Slater type*, but this required heavy computations. In 1969, Warren Hehre, Ross Stewart and John Pople published the article [86], in which these expansions were replaced by Gaussian functions, which made computing certain integrals much simpler. The coefficients in the Gauss expansion were computed using the least square method.

Right after this publication, Pople wrote a program called **Gaussian70** for computation based on new ideas. This program system was then expanded to many more versions including more general and more refined techniques, and today there are 19 versions, the latest being Gaussian09. Few program systems have been so dominating in such a broad scientific area as this one. In 1998, Pople received the Nobel prize together with Walter Kohn, and the motivation was: *"for his development of computational methods in quantum chemistry"*.

Even if the basic technique in the Gaussian system can be used for any type of molecules, it is in practice impossible for large ones. Reactions occur on such a small scale in both space and time that the computation becomes too much, even for the fastest computers of today. Classical Newton theory on the other hand can handle large molecules, but is unable to describe reactions on the microscale.

One possibility to overcome this difficulty is to combine the two techniques. This approach was taken by three distinguished researchers: Martin Karplus, Michael Levitt and Arieh Warshel. In the early 1970s they all started a combined approach, part of the work in cooperation. In 2013 they shared the Nobel Prize with motivation *"for the development of multiscale models for complex chemical systems"*. We have discussed **multiscale methods** in Sect. 5.9, and the methods used by the Nobel prize winners are closely related to those methods. However, here we have a special problem requiring special but very similar techniques.

6.3.3   Computational Engineering

Most modern aircraft are constructed after thorough computer simulation when it comes to the geometric design of the body (and follow up with wind tunnel tests). The form of the wings is of special importance since the wings provide the lift to get heavy aircraft to leave the ground and then fly with fuel consumption as small as possible. The principle for flying is that the pressure on top of the wing is lower than the pressure below, thereby creating a lift, and this is achieved by forming the wing so that the air speed is higher on the upper surface. Passenger aircrafts operate at a cruising speed around 900 km/h which is close to the speed of sound. The air speed on top of the wing becomes supersonic, and this in turn means that a shock sits on the top, i.e., there is a discontinuity in the pressure. This means that there is an extra force acting in the backward direction like a break acting on the aircraft. The main problem is to design the wing so that the troublesome shock is made as weak as possible.

No matter what kind of wing is shaped, there is a sharp bend in the geometry at the leading and the trailing edge. The computational grid must be very fine, and this is troublesome since any explicit difference method must use very small time-steps for the time-dependent PDE.

Antony Jameson is the best known scientist and engineer when it comes to computing the flow around an airfoil (a two-dimensional cut of the wing) or the full three-dimensional wing. The most basic problem is the steady state computation corresponding to flying at a constant speed. Jameson developed such methods based on a discretization in space, and the use of a **Runge–Kutta difference method** in time. Since accuracy in the time direction is not required, the time-step can be chosen locally fitted to the local space-step. Actually, the coefficients in the Runge–Kutta type scheme can be chosen in such a way that the stability domain in the complex plane becomes as large as possible, which results in fast convergence to steady state. Jameson also used a **multigrid method** for accelerating the convergence, sometimes in connection with implicit time-stepping methods.

The Euler equations are hyperbolic which is the reason for the presence of the shock. In order to avoid the troublesome oscillations occurring around the shock, **artificial viscosity** is introduced, see Sect. 4.3.5.

The design of wings and other parts of an aircraft leading to low fuel consumption represents the most important problems in engineering. The development of computational methods has had a considerable impact on advancing aircraft construction.

Antony Jameson
(Photographer P. Birken.
Permission for publication
given by A. Jameson)

References

1. A. Abdulle, W. E, B. Engquist, W. Ren, E. Vanden-Eijnden, The heterogeneous multiscale method. Acta Numer. **21**, 1–87 (2012)
2. J.H. Argyris, Energy theorems and structural analysis, part 1. Aircraft Eng. **26**, 383–387 (1954)
3. W.E. Arnoldi, The principle of minimized iterations in the solution of the matrix eigenvalue problem. Quart. Appl. Math. **9**, 17–29 (1951)
4. G.A. Baker, Finite element methods for elliptic equations using nonconforming elements. Math. Comput. **31**, 45–59 (1977)
5. F. Bashforth, J.C. Adams, *An Attempt to Test the Theories of Capillary Action by Comparing the Theoretical and Measured Forms of Drops of Fluid, with an Explanation of the Method of Integration Employed in Constructing the Tables which Give the Theoretical Forms of Such Drops* (Cambridge University Press, Cambridge, 1883)
6. R. Beatson, L. Greengard, A short course on fast multipole methods, in *Proceedings of Wavelets, Multilevel Methods and Elliptic PDEs* (Oxford University Press, Oxford, 1997), pp. 1–37
7. M. Benzi, Preconditioning techniques for large linear systems: a survey. J. Comput. Phys. **182**, 418–477 (2002)
8. J. Berenger, A perfectly matched layer for the absorption of electromagnetic waves. J. Comput. Phys. **114**, 185–200 (1994)
9. G. Beyklin, R. Coifman, V. Rokhlin, Fast wavelet transforms and numerical algorithms I. Commun. Pure Appl. Math. **44**, 141–183 (1991)
10. F. Black, M. Scholes, The pricing of options and corporate liabilities. J. Polit. Econ. **81**, 637–654 (1973)
11. R.N. Bracewell, Strip integration in radio astronomy. Aust. J. Phys. **9**, 192–217 (1956)
12. A. Brandt, Multi-level adaptive technique (MLAT) for fast numerical solution to boundary value problems, in *Proceedings of 3rd International Conference on Numerical Methods in Fluid Mechanics*. Lecture Notes in Physics, vol. 18 (Springer, Berlin, 1973), pp. 82–89
13. A. Brandt, Multi-level adaptive solutions to boundary-value problems. Math. Comput. **31**, 333–390 (1977)
14. C. Brezinski, D. Tournès, *André-Louis Cholesky; Mathematician, Topographer and Army Officer* (Birkhäuser, Basel, 2014)
15. R. Bulirsch, J. Stoer, Fehlerabschätzungen und Extrapolation mit rationalen Funktionen bei Verfahren vom Richardson-Typus. Numer. Math. **6**, 413–427 (1964)
16. J.C. Butcher, Coefficients for the study of Runge–Kutta integration processes. J. Aust. Math. Soc. **3**, 185–201 (1963)

© Springer International Publishing AG, part of Springer Nature 2018

B. Gustafsson, *Scientific Computing*, Texts in Computational Science and Engineering 17, https://doi.org/10.1007/978-3-319-69847-2

17. J.C. Butcher, On the attainable order of Runge–Kutta methods. Math. Comput. **19**, 408–417 (1965)
18. J.C. Butcher, *Numerical Methods for Ordinary Differential Equations. Runge-Kutta and General Linear Methods* (Wiley, Hoboken, 2003)
19. L. Cesari, Sulla risoluzione dei sistemi di equazioni lineari per approssimazioni successive. Atti Accad. Nazionale Lincei R. Classe Sci. Fis. Mat. Nat. **25**, 422 (1937)
20. J.-L. Chabert (ed.), *A History of Algorithms; From the Pebble to the Microchip* (Springer, Berlin, 1999)
21. J.G. Charney, R. Fjörtoft, J. von Neumann, Numerical integration of the barotropic vorticity equation. Tellus **2**, 237–254 (1950)
22. R.W. Clough, The finite element method in plane stress analysis, in *Proceedings of 2nd ASCE Conference on Electronic Computation, Pittsburgh, PA* (1960)
23. R.W. Clough, Original formulation of the finite element method, in *Proceedings of ASCE Structures Congress Session on Computer Utilization in Structural Engineering, San Francisco* (1989), pp. 1–10
24. J.W. Cooley, J.W. Tukey, An algorithm for the machine calculation of complex Fourier series. Math. Comput. **19**, 297–301 (1965)
25. A. Cormack, Representation of a function by its line integrals, with some radiological applications. J. Appl. Phys. **34**, 2722–2726 (1963)
26. A. Cormack, Representation of a function by its line integrals, with some radiological applications. II. J. Appl. Phys. **35**, 2908–2912 (1964)
27. R. Courant, Variational methods for the solution of problems of equilibrium and vibrations. Bull. Am. Math. Soc. **49**, 1–23 (1943)
28. R. Courant, K.O. Friedrichs, H. Levy, Über die partiellen Differenzengleichungen in der Mathematischen Physik. Math. Ann. **100**, 32–74 (1928)
29. R. Courant, E. Isaacson, M. Rees, On the solution of nonlinear hyperbolic differential equations by finite differences. Commun. Pure Appl. Math. **5**, 243–255 (1952)
30. R. Courant, K.O. Friedrichs, H. Levy, On the partial difference equations of mathematical physics. IBM J. Res. Dev. **11**, 215–234 (1967)
31. C.F. Curtiss, J.O. Hirschfelder, Integration of stiff equations. Proc. Nat. Acad. Sci. USA **638**, 235–243 (1952)
32. G. Dahlquist, Convergence and stability in the numerical integration of ordinary differential equations. Mat. Scand. **4**, 33–53 (1956)
33. G. Dahlquist, A special stability problem for linear multistep methods. BIT **3**, 27–43 (1963)
34. G.B. Dantzig, *Linear Programming and Extensions* (Princeton University Press, Princeton, 1963)
35. I. Daubechies, Orthonormal bases of compactly supported wavelets. Commun. Pure Appl. Math. **41**, 909–996 (1988)
36. I. Daubechies, *Ten Lectures on Wavelets* (SIAM, Philadelphia, 1992)
37. C. de Boor, On calculating with B-splines. J. Approx. Theory **6**, 50–62 (1972)
38. C. de Boor, *A Practical Guide to Splines* (Springer, Berlin, 1978)
39. P. Debye, Näherungsformeln für die Zylinderfunktionen für große Werte des Arguments und unbeschränkt veränderliche Werte des Index. Math. Ann. **67**, 535–558 (1904)
40. B. Delaunay, Sur la sphère vide. Otdelenie Matematicheskikh i Estestvennykh Nauk **7**, 793–800 (1934)
41. J. Dongarra, C. Moler, *LINPACK: Users' Guide* (SIAM, Philadelphia, 1979)
42. J. Douglas, On the numerical integration of $\partial^2 u/\partial x^2 + \partial^2 u/\partial y^2 = \partial u/\partial t$ by implicit methods. J. Soc. Ind. Appl. Math. **3**, 42–65 (1955)
43. J. Douglas, H.H. Rachford, On the numerical solution of heat conduction problems in two and three space variables. Trans. Am. Math. Soc. **82**, 421–439 (1956)
44. P.G. Drazin (ed.), *Collected Papers of Lewis Fry Richardson*, vol. 1 (Cambridge University Press, Cambridge, 1993)

45. M. Dryja, An algorithm with a capacitance matrix for a variational-difference scheme, in *Variational-Difference Methods in Mathematical Physics*, ed. G.I. Marchuk (Novosibirsk, USSR Academy of Sciences, 1981), pp. 63–73
46. M. Dryja, O. Widlund, Some domain decomposition algorithms for elliptic problems, in *Proceedings of Iterative Methods for Large Linear Systems, Austin 1988*, ed. by L. Hayes, D. Kincaid (Academic Press, New York, 1989), pp. 273–291
47. M. Engeli, T. Ginsburg, H. Rutishauser, H. Stiefel, *Refined Iterative Methods for Computation of the Solution and the Eigenvalues of Self-Adjoint Boundary Value Problems* (Birkhäuser, Basel, 1959)
48. B. Engquist (ed.), *Encyclopedia of Applied and Computational Mathematics* (Springer, Berlin, 2015)
49. B. Engquist, A. Majda, Absorbing boundary conditions for the numerical simulation of waves. Math. Comput. **31**, 629–651 (1977)
50. L. Euler, *Institutionum Calculi Integralis*, chap.7 (Impensis Academiae Imperialis Scientiarum, Petropoli, 1768)
51. R.P. Fedorenko, A relaxation method for solving elliptic difference equations. Z. Vycisl. Mat. i Mat. Fiz. **1**, 922–927 (1961) (Russian)
52. R.P. Fedorenko, The speed of convergence of one iterative process. USSR Comput. Math. Math. Phys. **4**, 227–235 (1964)
53. J. Fourier, *Théorie analytique de la chaleur* (Firmin Didot Père et Fils, Paris, 1822)
54. J. Fourier, Histoire de l'Académie, partie mathématique (1824), in *Mémoires de l'Académie des Sciences de l'Institut de France*, vol. 7 (Gauthier–Villars, Paris, 1827)
55. L. Fox, E.T. Goodwin, Some new methods for the numerical integration of ordinary differential equations. Proc. Camb. Phil. Soc. **45**, 373–388 (1949)
56. J.G.F. Francis, The qr transformation, I. Computer J. **4**, 265–271 (1961)
57. J.G.F. Francis, The qr transformation, II. Computer J. **4**, 332–345 (1962)
58. S. Frankel, Convergence rates of iterative treatments of partial differential equations. Math. Tables Aids Comput. **4**, 65–75 (1950)
59. D. Gabor, Theory of communication. J. IEEE **93**, 429–457 (1946)
60. B.G. Galerkin, Beams and plates. Using series for some problems in the elastic equilibrium of rods and plates. Eng. Bull. (Vestnik Inzhenerov) **19**, 897–908 (1915) (in Russian)
61. M.J. Gander, G. Wanner, From Euler, Ritz, and Galerkin to modern computing. SIAM Rev. **54**, 627–666 (2012)
62. C.F. Gauss, *Theoria Motus Corporum Coelestium in sectionibus conicis solem ambientium* (Perthes, Hamburg, 1809)
63. C.F. Gauss, Methodus nova integralium valores per approximationem inveniendi. Commentationens Sociatatis regiae scientarium Göttingensis recentiores **3**, 39–76 (1814)
64. C.F. Gauss, *Nachlass: theoria interpolationis methodo nova tractata* (Königliche Gesellschaft der Wissenschaften, Göttingen, 1866), pp. 265–330
65. C.W. Gear, The numerical integration of ordinary differential equations. Math. Comput. **21**, 146–156 (1967)
66. C.W. Gear, *Numerical Initial Value Problems in Ordinary Differential Equations* (Prentice Hall, Englewood Cliffs, 1971)
67. C.W. Gear, Numerical solution of ordinary differential equations; is there anything left to do? SIAM Rev. **23**, 10–24 (1981)
68. J.W. Gibbs, Fourier series. Nature **59**, 78–79 (1898)
69. J.W. Gibbs, Fourier series. Nature **59**, 233–234 (1899)
70. S.K Godunov, Difference methods for the numerical computation of discontinuous solutions of equations of hydrodynamics. Mat. Sb. N. S. **47**(89), 3, 271–306 (1959)
71. S.K Godunov, V.S. Ryabenkii, Special criteria for stability for boundary-value problems for non-self-adjoint difference equations. Uspekhi Mat. Nauk. **18**, 3–14 (1963)
72. H.H. Goldstine, *A History of Numerical Analysis from the 16th Through the 19th Century* (Springer, Berlin, 1977)

73. G.H. Golub, W. Kahan, Calculating the singular values and pseudo-inverse of a matrix. SIAM J. Numer. Anal. **2**, 205–224 (1965)
74. G.H. Golub, C. Reinsch, Singular value decomposition and least squares solution. Numer. Math. **14**, 403–420 (1970)
75. G.H. Golub, J.H. Welsch, Calculation of Gauss quadrature rules. Math. Comput. **23**, 221–230 (1969)
76. D. Gottlieb, The stability of pseudospectral–Chebyshev methods. Math. Comput. **36**, 107–118 (1981)
77. D. Gottlieb, B. Gustafsson, P. Forssén, On the direct Fourier method for computer tomography. IEEE Trans. Med. Imag. **19**, 223–232 (2000)
78. W.B. Gragg, On extrapolation algorithms for ordinary initial value problems. SIAM. J. Numer. Anal. **2**, 384–403 (1965)
79. L. Greengard, V. Rokhlin, A fast algorithm for particle simulation. J. Comput. Phys. **73**, 325–348 (1987)
80. A. Grossman, J. Morlet, Decomposition of Hardy functions into square integrable wavelets of constant shape. SIAM J. Math. Anal. **15**, 2473–2479 (1984)
81. B. Gustafsson, L. Hemmingsson-Frändén, A fast domain decomposition high order Poisson solver. J. Sci. Comput. **14**, 223–243 (1999)
82. B. Gustafsson, H.-O. Kreiss, A. Sundström, Stability theory of difference approximations for mixed initial boundary value problems II. Math. Comput. **26**, 649–686 (1972)
83. A. Haar, Zur Theorie der orthogonalen Funktionensysteme, (Erste Mitteilung). Math. Ann. **2669**, 331–371 (1910)
84. W. Hackbusch, *Multi-Grid Methods and Applications* (Springer, Berlin, 1985)
85. E. Hairer, G. Wanner, *Solving Ordinary Differential Equations II: Stiff and Differential–Algebraic Problems*, Springer Series in Computational Mathematics, 2nd edn. (Springer, Berlin, 1996)
86. W.J. Hehre, R.F. Stewart, J.A. Pople, Self-consistent molecular-orbital methods. I. Use of Gaussian expansions of Slater-type atomic orbitals. J. Chem. Phys. **51**, 2657–2664 (1969)
87. M.T. Heideman, D.H. Johnson, C.S. Burrus, Gauss and the history of the fast Fourier transform. Arch. Hist. Exact Sci. **34**, 265–277 (1985)
88. P. Henrici, *Discrete Variable Methods in Ordinary Differential Equations* (Wiley, Hoboken, 1962)
89. M. Hestenes, E. Stiefel, Methods of conjugate gradients for solving linear systems. J. Res. Natl. Bur. Stand. **49**, 409–438 (1952)
90. J.S. Hesthaven, T. Warburton, *Nodal Discontinuous Galerkin Methods Algorithms, Analysis and Applications*. Texts in Applied Mathematics, vol. 54 (Springer, Berlin, 2008)
91. J.S. Hesthaven, S. Gottlieb, D. Gottlieb, *Spectral Methods for Time-Dependent Problems*. Cambridge Monographs on Applied and Computational Mathematics (Cambridge University Press, Cambridge, 2007)
92. K. Heun, Neue Methode zur Approximativen Integration der Differentialgleichungen einer unabhängigen Veränderlichen (New methods for approximate integration of differential equations of one independent variable). Zeit. Math. Phys. **45**, 23–38 (1900)
93. G.N. Hounsfield, Computerized transverse axial scanning (tomography): part I. Description of system. Br. J. Radiol. **46**, 1016–1022 (1973)
94. C.G.J. Jacobi, Über eine neue Auflösungsart der bei der Methode der kleinsten Quadrate vorkommenden linearen Gleichungen. Astron. Nachrichten **22**, 297–298 (1845)
95. E.J. Kansa, Multiquadratics–a scattered data approximation scheme with applications to computational fluid dynamics II. Solutions to parabolic, hyperbolic and elliptic partial differential equations. Comput. Math. Appl. **19**, 147–161 (1990)
96. L. Kantorovich, *The Mathematical Method of Production Planning and Organization* (Leningrad University Press, Leningrad, 1939)
97. H.-O. Kreiss, Über die Lösung von Anfangswertaufgaben für partielle Differential-gleichungen. Trans. R. Inst. Technol. **166** (1960)

98. H.-O. Kreiss, Über die Stabilitätsdefinition für Differentialgleichungen die partielle Differentialgleichungen approximieren. BIT **2**, 153–181 (1962)
99. H.-O. Kreiss, On difference approximations of the dissipative type for hyperbolic differential equations. Commun. Pure Appl. Math. **17**, 335–353 (1964)
100. H.-O. Kreiss, Stability theory for difference approximations of mixed initial boundary value problems. I. Math. Comput. **22**, 703–714 (1968)
101. H.-O. Kreiss, Initial boundary value problems for hyperbolic systems. Commun. Pure Appl. Math. **23**, 277–298 (1970)
102. H.-O. Kreiss, J. Oliger, Comparison of accurate methods for the integration of hyperbolic problems. Tellus **24**, 199–215 (1972)
103. A.N. Krylov, On the numerical solution of the equation by which in technical questions frequencies of small oscillations of material systems are determined. Izvestiya Akademii Nauk SSSR, Otdelenie Matematicheskikh i Estestvennykh Nauk **7**, 491–539 (1931)
104. V.N. Kublanovskaya, On some algorithms for the solution of the complete eigenvalue problem. USSR Comput. Math. Math. Phys. **1**, 555–570 (1961)
105. C. Lanczos, Trigonometric interpolation of empirical and analytical functions. J. Math. Phys. **17**, 123–199 (1938)
106. C. Lanczos, Solutions of systems of linear equations by minimized iterations. J. Res. Natl. Bur. Stand. **49**, 33–53 (1952)
107. E. Larsson, B. Fornberg, A numerical study of some radial basis function based solution methods for elliptic PDEs. Comput. Math. Appl. **46**, 891–902 (2003)
108. E. Larsson, S.M. Gomes, A. Heryudono, A. Safdari-Vaighani, Radial basis function methods in computational finance, in *Proceedings of CMMSE13, Almeria, Spain* (2013), 12 pp.
109. P.D. Lax, Weak solutions of nonlinear hyperbolic equations and their numerical computation. Commun. Pure Appl. Math. **7**, 159–193 (1954)
110. P.D. Lax, Hyperbolic systems of conservation laws, II. Commun. Pure Appl. Math. **10**, 537–566 (1957)
111. P.D. Lax, John von Neumann: the early years, the years at Los Alamos, and the road to computing. SIAM News **38**(2) (2005)
112. P.D. Lax, R.D. Richtmyer, Survey of the stability of linear finite difference equations. Commun. Pure Appl. Math. **9**, 267–293 (1956)
113. A.-M. Legendre, *Nouvelles méthodes pour la détermination des orbites des comètes* (New methods for the determination of the orbits of comets) (Didot, Paris, 1805)
114. S.J. Leon, Å. Björck, W. Gander, Gram–Schmidt orthogonalization: 100 years and more. Numer. Linear Algebra Appl. **20**, 492–532 (2013)
115. P.-L. Lions, On the Schwarz alternating method I, in *First International Symposium on Domain Decomposition Methods for Partial Differential Equations*, ed. by R. Glowinski, G. Golub, G.A. Meurant, J. Périaux (1988), pp. 1–42
116. P.-L. Lions, *Mathematical Topics in Fluid Mechanics, Vol. 1: Compressible Models* (Oxford University Press, New York, 2013)
117. P.-L. Lions, *Mathematical Topics in Fluid Mechanics, Vol. II: Incompressible Models* (Oxford University Press, New York, 2013)
118. S.H. Lo, A new mesh generation scheme for arbitrary planar domains. Int. J. Numer. Math. Eng. **21**, 1403–1426 (1985)
119. N. Metropolis, S. Ulam, The Monte Carlo method. J. Am. Stat. Assoc. **44**, 335–341 (1949)
120. S.G. Mikhlin, On the Schwarz algorithm. Doklady Akademii Nauk SSSR **177**, 569–571 (1951)
121. E.H. Moore, On the reciprocal of the general algebraic matrix. Bull. Am. Math. Soc. **26**, 394–395 (1920)
122. F.R. Moulton, *New Methods in Exterior Ballistics* (University of Chicago Press, Chicago, 1926)
123. I. Newton, Methodus differentialis, in *Analysis Per Quantitatum Series, Fluxiones, ac Differentias: cum Enumeratione Linearum Tertii Ordinis, London* (1711)

124. E.J. Nyström, Über die numerische Integration von Differentialgleichungen. Acta Soc. Sci. Fennicae **50**, 1–55 (1925)

125. S. Orszag, Numerical methods for the simulation of turbulence. Phys. Fluids Suppl. II **12**, 250–257 (1969)

126. S.A. Orszag, Numerical simulation of incompressible flows within simple boundaries I. Galerkin (spectral) representations. Stud. Appl. Math. **50**, 293–327 (1971)

127. S.A. Orszag, Comparison of pseudospectral and spectral approximation. Stud. Appl. Math. **51**, 253–259 (1972)

128. S. Osher, J.A. Sethian, Fronts propagating with curvature dependent speed: algorithms based on Hamilton–Jacobi formulations. J. Comput. Phys. **79**, 12–49 (1988)

129. S. Osher, C.W. Shu, High order essentially non-oscillatory schemes for Hamilton–Jacobi equations. SIAM J. Numer. Anal. **28**, 907–922 (1991)

130. A.T. Patera, A spectral element method for fluid dynamics: laminar flow in a channel expansion. J. Comput. Phys. **554**, 468–488 (1984)

131. D.W. Peaceman, H.H. Rachford, The numerical solution of parabolic and elliptic differential equations. J. Soc. Ind. Appl. Math. **3**, 28–41 (1955)

132. R. Penrose, A generalized inverse for matrices. Proc. Camb. Philos. Soc. **51**, 406–413 (1955)

133. J. Peraire, M. Vahdati, K. Morgan, O. Zienkiewicz, Adaptive remeshing for compressible flow computations. J. Comput. Phys. **72**, 449–466 (1987)

134. V. Pereyra, On improving an approximate solution of a functional equation by deferred corrections. Numer. Math. **8**, 376–391 (1966)

135. H. Poincaré, *Les méthodes nouvelles de la mécanique céleste*, vol. 1 (Gauthier–Villars, Paris, 1892)

136. B. Rayleigh, J.W. Strutt, *The Theory of Sound*, vol. II (MacMillan, London, 1896)

137. W.H. Reed, T.R. Hill, Triangular mesh methods for the neutron transport equation. Los Alamos Scientific Laboratory, Tech. Report LA-UR-73-479, 1973

138. L.F. Richardson, The approximate arithmetical solution by finite differences of physical problems involving differential equations, with an application to the stresses in a masonry dam. Phil. Trans. R. Soc. A **210**, 307–357 (1911)

139. W. Ritz, Über eine neue Methode zur Lösung gewisser Variationsprobleme der mathematischen Physik. J. Reine Angew. Math. **135**, 1–61 (1908)

140. W. Ritz, Theorie der Transversalschwingungen einer quadratischen Platte mit freien Rändern. Ann. Phys. **18**, 737–807 (1909)

141. V. Rokhlin, Rapid solution of integral equations of scattering theory in two dimensions. J. Comput. Phys. **60**, 187–207 (1985)

142. C. Runge, Über die numerische Auflösung von Differentialgleichungen (on the numerical solution of differential equations. Math. Ann. **46**, 167–178 (1895)

143. C. Runge, Über empirische Funktionen und die Interpolation zwischen äquidistanten Ordinaten. Z. Math. Phys. **46**, 224–243 (1901)

144. H. Rutishauser, Über die Instabilität von Methoden zur Integration gewöhnlicher Differentialgleichungen. Z. Angew. Math. Phys. **3**, 65–74 (1952)

145. H. Rutishauser, Solution of eigenvalue problems with the lr-transformation. NBS Appl. Math. Ser. Phil. Trans. R. Soc. A **49**, 47–81 (1958)

146. Y. Saad, M.H. Schultz, GMRES: a generalized minimal residual algorithm for solving nonsymmetric linear systems. SIAM J. Sci. Stat. Comput. **7**, 856–869 (1986)

147. M. Saigey, *Problèmes d'arithmétique et exercises de calcul du second degré avec les solutions raisonnées* (Hachette, Paris, 1859)

148. J.B. Scarborough, *Numerical Mathematical Analysis* (The Johns Hopkins Press, Baltimore, 1930)

149. I.J. Schoenberg, Contributions to the problem of approximation of equidistant data by analytic functions. Quart. Appl. Math. **4**, 45–99, 112–141 (1946)

150. H.A. Schwarz, Über einen Grenzübergang durch alternierendes Verfahren. Vierteljahrsschrift der Naturforschenden Gesellschaft Zürich **15**, 272–286 (1870)

151. G.W. Stewart, On the early history of the singular value decomposition. SIAM Rev. **35**, 551–566 (1993)
152. G. Strang, Wavelets. Am. Sci. **82**, 250–255 (1994)
153. A. Toselli, O.B. Widlund, *Domain Decomposition Methods - Algorithms and Theory*. Springer Series in Computational Mathematics (Springer, Berlin, 2005)
154. R. Tsai, S. Osher, Level set methods and their applications in image science. Commun. Math. Sci. **1**, 1–20 (2003)
155. M.J. Turner, R.W. Clough, H.C. Martin, L.J. Topp, Stiffness and deflection analysis of complex structures. J. Aero. Sci. **23**, 805–823 (1956)
156. R. von Mises, H. Pollaczek-Geiringer, Praktische Verfahren der Gleichungsauflösung. ZAMM **9**, 152–164 (1929)
157. J. von Neumann, Proposal and analysis of a numerical method for the treatment of hydrodynamic shock problems. National Defense and Research Committee Report AM551, 1944
158. J. von Neumann, R. Richtmyer, A method for numerical calculation of hydrodynamic shocks. J. Appl. Phys. **21**, 232–237 (1950)
159. L. von Sydow et al. BENCHOP–the BENCHmarking project in option pricing. Int. J. Comput. Math. **92**, 2361–2379 (2015)
160. K. Weierstrass, Über die analytische Darstellbarkeit sogenannter willkürlicher Funktionen einer reellen Veränderlichen. *Sitzungsberichte der Königlich Preussischen Akademie der Wissenschaften zu Berlin*, II: Erste Mitteilung (Part 1) 633–639, Zweite Mitteilung (Part 2) 789–805, 1885
161. E. Weinan, B. Engquist, The heterogeneous multiscale methods. Commun. Math. Sci. **1**, 87–132 (2003)
162. E.T. Whittaker, G. Robinson, *The Calculus of Observations. A Treatise on Numerical Mathematics* (Blackie & Son, London, 1924)
163. O.C. Zienkiewicz, Y.K. Cheung, Finite elements in the solution of field problems. The Engineer **220**, 507–510 (1965)

Index

Editorial Policy

1. Textbooks on topics in the field of computational science and engineering will be considered. They should be written for courses in CSE education. Both graduate and undergraduate textbooks will be published in TCSE. Multidisciplinary topics and multidisciplinary teams of authors are especially welcome.

2. Format: Only works in English will be considered. For evaluation purposes, manuscripts may be submitted in print or electronic form, in the latter case, preferably as pdf- or zipped ps-files. Authors are requested to use the LaTeX style files available from Springer at: http://www.springer.com/authors/book+authors/helpdesk?SGWID=0-1723113-12-971304-0 (Click on ⟶ Templates ⟶ LaTeX ⟶ monographs)
Electronic material can be included if appropriate. Please contact the publisher.

3. Those considering a book which might be suitable for the series are strongly advised to contact the publisher or the series editors at an early stage.

General Remarks

Careful preparation of manuscripts will help keep production time short and ensure a satisfactory appearance of the finished book.

The following terms and conditions hold:

Regarding free copies and royalties, the standard terms for Springer mathematics textbooks hold. Please write to martin.peters@springer.com for details.

Authors are entitled to purchase further copies of their book and other Springer books for their personal use, at a discount of 33.3% directly from Springer-Verlag.

Series Editors

Timothy J. Barth
NASA Ames Research Center
NAS Division
Moffett Field, CA 94035, USA
barth@nas.nasa.gov

Michael Griebel
Institut für Numerische Simulation
der Universität Bonn
Wegelerstr. 6
53115 Bonn, Germany
griebel@ins.uni-bonn.de

David E. Keyes
Mathematical and Computer Sciences
and Engineering
King Abdullah University of Science
and Technology
P.O. Box 55455
Jeddah 21534, Saudi Arabia
david.keyes@kaust.edu.sa

and

Department of Applied Physics
and Applied Mathematics
Columbia University
500 W. 120 th Street
New York, NY 10027, USA
kd2112@columbia.edu

Risto M. Nieminen
Department of Applied Physics
Aalto University School of Science
and Technology
00076 Aalto, Finland
risto.nieminen@tkk.fi

Dirk Roose
Department of Computer Science
Katholieke Universiteit Leuven
Celestijnenlaan 200A
3001 Leuven-Heverlee, Belgium
dirk.roose@cs.kuleuven.be

Tamar Schlick
Department of Chemistry
and Courant Institute
of Mathematical Sciences
New York University
251 Mercer Street
New York, NY 10012, USA
schlick@nyu.edu

Editors for Computational Science
and Engineering at Springer:

For Lecture Notes in Computational
Science and Engineering
Jan Holland
jan.holland@springer.com

For Texts in Computational Science
and Engineering and
Monographs in Computational
Science and Engineering
Martin Peters
martin.peters@springer.com

Springer-Verlag
Mathematics Editorial
Tiergartenstr. 17
69121 Heidelberg
Germany

Texts in Computational Science and Engineering

For further information on these books please have a look at our mathematics catalogue at the following URL: www.springer.com/series/5151

Monographs in Computational Science and Engineering

For further information on this book, please have a look at our mathematics catalogue at the following URL: www.springer.com/series/7417

Lecture Notes in Computational Science and Engineering

24. T. Schlick, H.H. Gan (eds.), *Computational Methods for Macromolecules: Challenges and Applications.*

25. T.J. Barth, H. Deconinck (eds.), *Error Estimation and Adaptive Discretization Methods in Computational Fluid Dynamics.*

26. M. Griebel, M.A. Schweitzer (eds.), *Meshfree Methods for Partial Differential Equations.*

27. S. Müller, *Adaptive Multiscale Schemes for Conservation Laws.*

28. C. Carstensen, S. Funken, W. Hackbusch, R.H.W. Hoppe, P. Monk (eds.), *Computational Electromagnetics.*

29. M.A. Schweitzer, *A Parallel Multilevel Partition of Unity Method for Elliptic Partial Differential Equations.*

30. T. Biegler, O. Ghattas, M. Heinkenschloss, B. van Bloemen Waanders (eds.), *Large-Scale PDE-Constrained Optimization.*

31. M. Ainsworth, P. Davies, D. Duncan, P. Martin, B. Rynne (eds.), *Topics in Computational Wave Propagation.* Direct and Inverse Problems.

32. H. Emmerich, B. Nestler, M. Schreckenberg (eds.), *Interface and Transport Dynamics.* Computational Modelling.

33. H.P. Langtangen, A. Tveito (eds.), *Advanced Topics in Computational Partial Differential Equations.* Numerical Methods and Diffpack Programming.

34. V. John, *Large Eddy Simulation of Turbulent Incompressible Flows.* Analytical and Numerical Results for a Class of LES Models.

35. E. Bänsch (ed.), *Challenges in Scientific Computing - CISC 2002.*

36. B.N. Khoromskij, G. Wittum, *Numerical Solution of Elliptic Differential Equations by Reduction to the Interface.*

37. A. Iske, *Multiresolution Methods in Scattered Data Modelling.*

38. S.-I. Niculescu, K. Gu (eds.), *Advances in Time-Delay Systems.*

39. S. Attinger, P. Koumoutsakos (eds.), *Multiscale Modelling and Simulation.*

40. R. Kornhuber, R. Hoppe, J. Périaux, O. Pironneau, O. Wildlund, J. Xu (eds.), *Domain Decomposition Methods in Science and Engineering.*

41. T. Plewa, T. Linde, V.G. Weirs (eds.), *Adaptive Mesh Refinement – Theory and Applications.*

42. A. Schmidt, K.G. Siebert, *Design of Adaptive Finite Element Software.* The Finite Element Toolbox ALBERTA.

43. M. Griebel, M.A. Schweitzer (eds.), *Meshfree Methods for Partial Differential Equations II.*

44. B. Engquist, P. Lötstedt, O. Runborg (eds.), *Multiscale Methods in Science and Engineering.*

45. P. Benner, V. Mehrmann, D.C. Sorensen (eds.), *Dimension Reduction of Large-Scale Systems.*

46. D. Kressner, *Numerical Methods for General and Structured Eigenvalue Problems.*

47. A. Boriçi, A. Frommer, B. Joó, A. Kennedy, B. Pendleton (eds.), *QCD and Numerical Analysis III.*

48. F. Graziani (ed.), *Computational Methods in Transport.*

49. B. Leimkuhler, C. Chipot, R. Elber, A. Laaksonen, A. Mark, T. Schlick, C. Schütte, R. Skeel (eds.), *New Algorithms for Macromolecular Simulation.*

50. M. Bücker, G. Corliss, P. Hovland, U. Naumann, B. Norris (eds.), *Automatic Differentiation: Applications, Theory, and Implementations.*

51. A.M. Bruaset, A. Tveito (eds.), *Numerical Solution of Partial Differential Equations on Parallel Computers.*

52. K.H. Hoffmann, A. Meyer (eds.), *Parallel Algorithms and Cluster Computing.*

53. H.-J. Bungartz, M. Schäfer (eds.), *Fluid-Structure Interaction.*

54. J. Behrens, *Adaptive Atmospheric Modeling.*

55. O. Widlund, D. Keyes (eds.), *Domain Decomposition Methods in Science and Engineering XVI.*

56. S. Kassinos, C. Langer, G. Iaccarino, P. Moin (eds.), *Complex Effects in Large Eddy Simulations.*

57. M. Griebel, M.A Schweitzer (eds.), *Meshfree Methods for Partial Differential Equations III.*

58. A.N. Gorban, B. Kégl, D.C. Wunsch, A. Zinovyev (eds.), *Principal Manifolds for Data Visualization and Dimension Reduction.*

59. H. Ammari (ed.), *Modeling and Computations in Electromagnetics: A Volume Dedicated to Jean-Claude Nédélec.*

60. U. Langer, M. Discacciati, D. Keyes, O. Widlund, W. Zulehner (eds.), *Domain Decomposition Methods in Science and Engineering XVII.*

61. T. Mathew, *Domain Decomposition Methods for the Numerical Solution of Partial Differential Equations.*

62. F. Graziani (ed.), *Computational Methods in Transport: Verification and Validation.*

63. M. Bebendorf, *Hierarchical Matrices. A Means to Efficiently Solve Elliptic Boundary Value Problems.*

64. C.H. Bischof, H.M. Bücker, P. Hovland, U. Naumann, J. Utke (eds.), *Advances in Automatic Differentiation.*

65. M. Griebel, M.A. Schweitzer (eds.), *Meshfree Methods for Partial Differential Equations IV.*

66. B. Engquist, P. Lötstedt, O. Runborg (eds.), *Multiscale Modeling and Simulation in Science.*

67. I.H. Tuncer, Ü. Gülcat, D.R. Emerson, K. Matsuno (eds.), *Parallel Computational Fluid Dynamics 2007.*

68. S. Yip, T. Diaz de la Rubia (eds.), *Scientific Modeling and Simulations.*

69. A. Hegarty, N. Kopteva, E. O'Riordan, M. Stynes (eds.), *BAIL 2008 – Boundary and Interior Layers.*

70. M. Bercovier, M.J. Gander, R. Kornhuber, O. Widlund (eds.), *Domain Decomposition Methods in Science and Engineering XVIII.*

71. B. Koren, C. Vuik (eds.), *Advanced Computational Methods in Science and Engineering.*

72. M. Peters (ed.), *Computational Fluid Dynamics for Sport Simulation.*

73. H.-J. Bungartz, M. Mehl, M. Schäfer (eds.), *Fluid Structure Interaction II - Modelling, Simulation, Optimization.*

74. D. Tromeur-Dervout, G. Brenner, D.R. Emerson, J. Erhel (eds.), *Parallel Computational Fluid Dynamics 2008.*

75. A.N. Gorban, D. Roose (eds.), *Coping with Complexity: Model Reduction and Data Analysis.*

76. J.S. Hesthaven, E.M. Rønquist (eds.), *Spectral and High Order Methods for Partial Differential Equations.*

77. M. Holtz, *Sparse Grid Quadrature in High Dimensions with Applications in Finance and Insurance.*

78. Y. Huang, R. Kornhuber, O.Widlund, J. Xu (eds.), *Domain Decomposition Methods in Science and Engineering XIX.*

79. M. Griebel, M.A. Schweitzer (eds.), *Meshfree Methods for Partial Differential Equations V.*

80. P.H. Lauritzen, C. Jablonowski, M.A. Taylor, R.D. Nair (eds.), *Numerical Techniques for Global Atmospheric Models.*

81. C. Clavero, J.L. Gracia, F.J. Lisbona (eds.), *BAIL 2010 – Boundary and Interior Layers, Computational and Asymptotic Methods.*

82. B. Engquist, O. Runborg, Y.R. Tsai (eds.), *Numerical Analysis and Multiscale Computations.*

83. I.G. Graham, T.Y. Hou, O. Lakkis, R. Scheichl (eds.), *Numerical Analysis of Multiscale Problems.*

84. A. Logg, K.-A. Mardal, G. Wells (eds.), *Automated Solution of Differential Equations by the Finite Element Method.*

85. J. Blowey, M. Jensen (eds.), *Frontiers in Numerical Analysis - Durham 2010.*

86. O. Kolditz, U.-J. Gorke, H. Shao, W. Wang (eds.), *Thermo-Hydro-Mechanical-Chemical Processes in Fractured Porous Media - Benchmarks and Examples.*

87. S. Forth, P. Hovland, E. Phipps, J. Utke, A. Walther (eds.), *Recent Advances in Algorithmic Differentiation.*

88. J. Garcke, M. Griebel (eds.), *Sparse Grids and Applications.*

89. M. Griebel, M.A. Schweitzer (eds.), *Meshfree Methods for Partial Differential Equations VI.*

90. C. Pechstein, *Finite and Boundary Element Tearing and Interconnecting Solvers for Multiscale Problems.*

91. R. Bank, M. Holst, O. Widlund, J. Xu (eds.), *Domain Decomposition Methods in Science and Engineering XX.*

92. H. Bijl, D. Lucor, S. Mishra, C. Schwab (eds.), *Uncertainty Quantification in Computational Fluid Dynamics.*

93. M. Bader, H.-J. Bungartz, T. Weinzierl (eds.), *Advanced Computing.*

94. M. Ehrhardt, T. Koprucki (eds.), *Advanced Mathematical Models and Numerical Techniques for Multi-Band Effective Mass Approximations.*

95. M. Azaïez, H. El Fekih, J.S. Hesthaven (eds.), *Spectral and High Order Methods for Partial Differential Equations ICOSAHOM 2012.*

96. F. Graziani, M.P. Desjarlais, R. Redmer, S.B. Trickey (eds.), *Frontiers and Challenges in Warm Dense Matter.*

97. J. Garcke, D. Pflüger (eds.), *Sparse Grids and Applications – Munich 2012.*

98. J. Erhel, M. Gander, L. Halpern, G. Pichot, T. Sassi, O. Widlund (eds.), *Domain Decomposition Methods in Science and Engineering XXI.*

99. R. Abgrall, H. Beaugendre, P.M. Congedo, C. Dobrzynski, V. Perrier, M. Ricchiuto (eds.), *High Order Nonlinear Numerical Methods for Evolutionary PDEs - HONOM 2013.*

100. M. Griebel, M.A. Schweitzer (eds.), *Meshfree Methods for Partial Differential Equations VII.*

101. R. Hoppe (ed.), *Optimization with PDE Constraints - OPTPDE 2014.*

102. S. Dahlke, W. Dahmen, M. Griebel, W. Hackbusch, K. Ritter, R. Schneider, C. Schwab, H. Yserentant (eds.), *Extraction of Quantifiable Information from Complex Systems.*

103. A. Abdulle, S. Deparis, D. Kressner, F. Nobile, M. Picasso (eds.), *Numerical Mathematics and Advanced Applications - ENUMATH 2013.*

104. T. Dickopf, M.J. Gander, L. Halpern, R. Krause, L.F. Pavarino (eds.), *Domain Decomposition Methods in Science and Engineering XXII.*

105. M. Mehl, M. Bischoff, M. Schäfer (eds.), *Recent Trends in Computational Engineering - CE2014.* Optimization, Uncertainty, Parallel Algorithms, Coupled and Complex Problems.

106. R.M. Kirby, M. Berzins, J.S. Hesthaven (eds.), *Spectral and High Order Methods for Partial Differential Equations - ICOSAHOM'14.*

107. B. Jüttler, B. Simeon (eds.), *Isogeometric Analysis and Applications 2014.*

108. P. Knobloch (ed.), *Boundary and Interior Layers, Computational and Asymptotic Methods – BAIL 2014.*

109. J. Garcke, D. Pflüger (eds.), *Sparse Grids and Applications – Stuttgart 2014.*

110. H. P. Langtangen, *Finite Difference Computing with Exponential Decay Models.*

111. A. Tveito, G.T. Lines, *Computing Characterizations of Drugs for Ion Channels and Receptors Using Markov Models.*

112. B. Karazösen, M. Manguoğlu, M. Tezer-Sezgin, S. Göktepe, Ö. Uğur (eds.), *Numerical Mathematics and Advanced Applications - ENUMATH 2015.*

113. H.-J. Bungartz, P. Neumann, W.E. Nagel (eds.), *Software for Exascale Computing - SPPEXA 2013-2015.*

114. G.R. Barrenechea, F. Brezzi, A. Cangiani, E.H. Georgoulis (eds.), *Building Bridges: Connections and Challenges in Modern Approaches to Numerical Partial Differential Equations.*

115. M. Griebel, M.A. Schweitzer (eds.), *Meshfree Methods for Partial Differential Equations VIII.*

116. C.-O. Lee, X.-C. Cai, D.E. Keyes, H.H. Kim, A. Klawonn, E.-J. Park, O.B. Widlund (eds.), *Domain Decomposition Methods in Science and Engineering XXIII.*

117. T. Sakurai, S. Zhang, T. Imamura, Y. Yusaku, K. Yoshinobu, H. Takeo (eds.), *Eigenvalue Problems: Algorithms, Software and Applications, in Petascale Computing.* EPASA 2015, Tsukuba, Japan, September 2015.

118. T. Richter (ed.), *Fluid-structure Interactions.* Models, Analysis and Finite Elements.

119. M.L. Bittencourt, N.A. Dumont, J.S. Hesthaven (eds.), *Spectral and High Order Methods for Partial Differential Equations ICOSAHOM 2016.*

120. Z. Huang, M. Stynes, Z. Zhang (eds.), *Boundary and Interior Layers, Computational and Asymptotic Methods BAIL 2016.*

121. S.P.A. Bordas, E.N. Burman, M.G. Larson, M.A. Olshanskii (eds.), *Geometrically Unfitted Finite Element Methods and Applications.* Proceedings of the UCL Workshop 2016.

For further information on these books please have a look at our mathematics catalogue at the following URL: www.springer.com/series/3527

Printed in the United States
By Bookmasters